Introductory Stochastic Analysis for Finance and Insurance

Introductory Stochastic Analysis for Finance and Insurance

X. Sheldon Lin

University of Toronto
Department of Statistics
Toronto, Ontario, Canada

A JOHN WILEY & SONS, INC., PUBLICATION

Published by John Wiley & Sons, Inc., Hoboken, New Jersey.
Published simultaneously in Canada.

For general information on our other products and services or for technical support, please contact our Customer Care Department within the United States at (800) 762-2974, outside the United States at (317) 572-3993 or fax (317) 572-4002.

Wiley also publishes its books in a variety of electronic formats. Some content that appears in print may not be available in electronic formats. For more information about Wiley products, visit our web site at www.wiley.com.

Library of Congress Cataloging-in-Publication Data is available.

ISBN-13 978-0-471-71642-6
ISBN-10 0-471-71642-1

10 9 8 7 6 5 4 3

To my family

CONTENTS

LIST OF FIGURES

LIST OF TABLES

PREFACE

The aim of this book is to provide basic stochastic analysis techniques for mathematical finance. It is intended for those who want to study mathematical finance but have a limited background in probability theory and stochastic analysis. The writing of the book started seven years ago after I taught mathematical finance to graduate students and practitioners for several years. Most of my students have a Master's degree in science or engineering but had only taken one or two entry level probability courses prior to my course. My initial approach for teaching such a course was to focus on financial models and aspects of modelling techniques. I soon found that the lack of the sound knowledge in stochastic analysis often prevented students from fully understanding and mastering financial models, and the implementation of these models. I ended up spending much of the class time teaching stochastic analysis instead. I feel from that experience that a short course on stochastic analysis with an emphasis on techniques and methods used in financial modelling would greatly help students with a limited probability background to better study mathematical finance. This book is an attempt to meet such a need. To serve that purpose, I have written this book to be self-contained and to use many well-known stochastic models in finance for illustration. My hope is that after reading the book, readers are able to understand the technical aspects of most stochastic models in finance and are able to use this book as a stepping stone to learn more advanced topics in stochastic analysis for mathematical finance.

The development of this book has received a financial support from the Finance Research Committee and the Committee on Knowledge Extension Research of the Society of Actuaries. I am very grateful to the committees for the financial support and in particular to the committees' members: Mark Abbott at the Guardian Life, Sarah Christiansen at the SMC Actuarial Consultants, Curtis Huntington at the University of Michigan, John Manistre at AEGON, and Steve Siegel at the Society of Actuaries for their assistance and enthusiasm on the project.

This book has undergone much revision over the past several years, and has been used in graduate courses at the University of Iowa, the University of Toronto and the University of Michigan. Many colleagues and graduate students have made valuable comments and suggestions over those years. In particular, Ralph Russo at the University of Iowa, Samuel Broverman at the University of Toronto and David Kausch at the University of Michigan have read early drafts of the book and have made numerous valuable suggestions. I am very grateful for their suggestions. I wish to thank my current and former students, Patrice Gaillardetz, Hansuck Lee, Christine Lim, Xiaoming Liu, Jerome Pansera, Serena Tiong, and Tao Wang for their comments and input. Special thanks also goes to to Steve Quigley and Susanne Steitz of John Wiley and Sons for their support of the project and their patience.

I am deeply indebted to Phelim Boyle at the University of Waterloo and Elias Shiu at the University of Iowa, who are not only my colleagues but also my mentors and great friends. It was Phelim who introduced me to the wonderful world of mathematical finance. I continue to learn from him in many areas of finance and insurance. I am very fortunate to have had the opportunity to work alongside Elias at the University of Iowa for several years. I have benefited greatly from my experience during those years.

Finally, I wish to thank my colleagues and friends Hans Gerber at the University of Lausanne, Gordon Willmot at the University of Waterloo and Hailiang Yang the University of Hong Kong for their encouragement, and my wife Feng and son Jason for their support.

X.S. LIN

Toronto, Canada
December, 2005

CHAPTER 1

INTRODUCTION

The theory of stochastic processes studies sequences of time-related random events, termed as stochastic processes, governed by probability laws. It provides mathematical tools to model and analyze these events in both qualitative and quantitative ways. Stochastic calculus concerns a specific class of stochastic processes that are stochastically integrable and are often expressed as solutions to stochastic differential equations. Applications of stochastic processes and stochastic calculus may be found in many disciplines such as physics, engineering, finance and insurance.

Early financial applications of stochastic processes date back to at least year 1900 when the French mathematician Louis Bachelier applied a special stochastic process called Brownian motion or a Wiener process to describe stock prices in his Ph.D. dissertation[1]. Significant financial applications of

---

[1] Brownian motion was named after the botanist Robert Brown who first observed the 'random movement' when studying movement of pollen grains in water in 1827. This process is sometime called a Wiener process in mathematical/physical sciences, as the mathematician Norbert Wiener, the founder of Cybernetics, applied a theoretical description of Brownian motion to quantum phenomena. Bachelier's work was largely forgotten until the Nobel Laureate Paul Samuelson brought it to the attention of the economics community in the 1960s (see Robert Merton's speech to the Bachelier Finance Society, June 2000).

stochastic processes and stochastic calculus in particular came in the late 1960s and early 1970s when a series of seminal research papers in option pricing were published by Fisher Black, Robert Merton and Myron Scholes, among others. In recent years, stochastic processes and stochastic calculus have been applied to a wide range of financial problems. Besides the valuation of options, they have been used to model interest rates and evaluate fixed income securities including interest rate futures, interest rate options, interest rate caps and floors, swaps and swaptions, and mortgage-backed securities (MBSs). Applications can also be found in portfolio analysis, foreign exchange analysis and, more recently, credit risk analysis.

Stochastic processes have been used in modeling insurance claims and pricing insurance products for many years. Swedish actuaries Filip Lundberg and Harald Cramér applied Poisson processes to the occurence of insurance claims in the early development of insurance risk theory from the 1900s through the 1950s. The techniques and methods used in insurance valuation are traditionally statistical rather than stochastic and they are adequate when the design of insurance and annuity products is simple and the interest rate environment is relatively stable. However, the recent changes in investment environments make stochastic analysis necessary. First of all, there are an increasing number of fixed income securities and interest rate derivatives available such as MBSs, Asset-backed securities (ABSs), swaps, and swaptions in financial markets. Although the availability of them gives insurers more flexibility to manage and hedge their investment portfolios, these securities are more sensitive to the fluctuation of interest rates and as a result the valuation of these securities may require more sophisticated stochastic models to effectively capture the interplay among them. Secondly, many insurance companies have recently introduced equity-linked insurance and annuity products. A common feature of these products is that they guarantee certain floor benefits and at the same time link their return to the performance of an external equity index such as the S&P 500. In other words, equity-linked products provide some participation in the equity market but limit the downside risks. These products include equity-indexed annuities (EIA), variable annuities (VA) and Universal Life (UL), to name a few. Thus the imbedded guarantees in equity-linked products are similar to certain derivative securities in the financial market. The valuation of these guarantees inevitably involves the use of modern option pricing theory that requires the use of stochastic processes and stochastic calculus. In Chapter 7, we describe in some detail the product design of an EIA and UL, and illustrate the application of stochastic calculus to the valuation of these two products.

The aim of this book is to introduce the basic theories of stochastic processes and stochastic calculus and to provide necessary tool kits for modeling and pricing in finance and insurance. When treated rigorously, the theory of stochastic processes and stochastic calculus involves the extensive use of fairly advanced mathematics such as measure theory. Our focus will however be on applications rather than on the theory, emphasizing intuition and computational aspects.

This book consists of seven chapters. In Chapter 2, probability theory is reviewed briefly. Unlike most introductory texts in probability theory, the focus here is on:

(i) the notion of information structure and how it relates to a probability space and a random variable. Since a stochastic process any be viewed as a collection of random variables at different times with different information structures, a good understanding of information structure aids our understanding of how a stochastic process evolves over time;

(ii) the notion of conditional probability and conditional expectation and their calculation. Conditional probability and conditional expectation are central to stochastic calculus. The applications of most results in stochastic calculus require the calculation of conditional probabilities and/or conditional expectations.

In Chapter 3, discrete-time stochastic processes are introduced. We consider in detail a special class of discrete-time stochastic processes known as random walks due to its importance in financial modeling. We also use a random walk to illustrate important concepts associated with stochastic processes, such as change of probability measures, martingales, stopping times and applications of the Optional Sampling Theorem. Discrete-time Markov chains are discussed briefly. Several applications of random walks in finance are discussed. These include binomial models for a stock price, option pricing and interest rate models.

Continuous-time stochastic processes are the subject of Chapter 4. Brownian motion is introduced as a limit of a sequence of random walks. The Reflection Principle, crucial to one's understanding of Brownian motion, and barrier hitting time distributions are discussed. We also provide an alternative approach using the Poisson process. Most actuaries are familiar with the Poisson process through its use in modeling aggregate claim processes. The notions of change of probability measures, martingales and stopping times, and the Optional Sampling Theorem are discussed again but in a continuous-time setting. It should be pointed out that there are many excellent books on stochastic processes for the reader to learn more on the subject. Two books particularly worth mentioning are *Introduction to Probability Models* and *Stochastic Processes,* both by Sheldon Ross.

Stochastic calculus is covered in Chapters 5 and 6. We present the most useful and simplest stochastic calculus: the Ito calculus, which was developed by the mathematician Kiyoshi Ito. The Ito calculus provides mathematical tools for a special class of continuous-time stochastic processes (often called Ito processes) that include Brownian motion. There are similarities between the stochastic calculus and the standard calculus taught usually as a second year undergraduate course[2], as the latter provides mathematical tools for deterministic functions. We can roughly think that a stochastic process is a function whose range is a collection of random variables. However, there are some fundamental differences between the Ito calculus and the standard calculus. As we will see later, an Ito process is usually nowhere differentiable and hence derivatives of an Ito process can not be defined properly. For this reason, stochastic calculus begins with stochastic integration.

Chapter 5 introduces a specific stochastic integral called the Ito integral as it is the most useful for financial applications. Tools for Ito integrals such as Ito's Lemma, a stochastic version of the Chain Rule, and integration by parts are established. Stochastic differential equations and Ito processes as solutions to stochastic differential equations are introduced. Ito's Lemma and integration by parts are the building blocks of stochastic calculus. They are used not only to derive other results in stochastic calculus, but as essential tools in solving stochastic differential equations. The topics covered in this chapter are necessary for financial applications. We emphasize the applications of Ito's Lemma and integration by parts, rather than their rigorous mathematical proofs. We discuss at length on how to solve stochastic differential equations and how to construct martingales associated with Ito processes. Many examples are given, including the famous Black-Scholes Option Pricing Formula.

Chapter 6 is devoted to advanced topics in stochastic calculus. The Feynman-Kac Formula, the Girsanov Theorem and complex barrier hitting times, as well as high-dimensional stochastic differential equations are discussed. The results in this chapter are as important as those of Chapter 5. In fact, results such as the Feynman-Kac Formula and the Girsanov Theorem are essential in derivative pricing. However, they are technically difficult and require deeper understanding of advanced mathematics and probability theory. The reader may wish to avoid this chapter when reading the book for the first time.

Chapter 7 presents several applications in insurance valuation. The first application involves the valuation of variable annuities, the most popular deferred annuity product sold in the North America. The approach developed for variable annuities is extended to valuate Equity-Indexed Annuities, another equity-linked product sold in the North America. The third application involves the valuation of guaranteed annuity options often associated pension policies. Finally in this chapter, a Universal Life is considered.

---

[2]Some people refer to standard calculus as deterministic calculus.

CHAPTER 2

OVERVIEW OF PROBABILITY THEORY

In this chapter, we recall briefly some concepts and results in probability theory. We begin with probability spaces and information structures in Section 2.1. In some introductory probability textbooks, information structure and its association with a probability space is not discussed thoroughly. However, it is important to gain a good understanding of information structure in order to study stochastic calculus. Section 2.2 discusses random variables and their properties. Several common random variables and their distributions are given. Multivariate distributions are introduced in Section 2.3. Examples involving the multinomial distribution and bivariate normal distribution are given in this section. We consider conditional probability in Section 2.4 and conditional expectation in Section 2.5. Conditional probability and conditional expectation are extremely important since many results in stochastic calculus involve these notions. Section 2.6 concerns the Central Limit Theorem which is the foundation for the normal approximation.

Since this book focuses on stochastic processes, we do not discuss extensively the notions and results in probability theory. For a more thorough study of the topics covered in this chapter we refer readers to Hassett and Stewart

(1999), Hogg and Craig (1978), Ross (1993), and Williams (1994). It is crucial to understand probability theory in order to master the materials in later chapters.

2.1 PROBABILITY SPACES AND INFORMATION STRUCTURES

Consider an experiment whose outcome is unpredictable. The set of all possible outcomes from the experiment is called a sample space or a state space, and is denoted by Ω. An event of the state space Ω is a collection of outcomes and thus mathematically a subset of the state space Ω. An information structure[1], denoted by $\mathcal{F}$ on the same state space is a collection of certain events. Those events represent only those that may be observed, depending on how the observation is made. Consequently, there could be more than one information structure on the same state space. An example will be given later in this section. In order to quantify the unpredictability of the events in $\mathcal{F}$, we assign a real number between 0 and 1 to each and call it the probability of that event. As a result, we obtain a real-valued function P on $\mathcal{F}$, called the probability measure. For a particular event in $\mathcal{F}$, the probabilty of the event represents the relative frequency of the event when the experiment is performed repeatedly. It is clear from above that an information structure has to be given before a probability measure is assigned.

As an example, we consider the experiment where a fair coin is tossed twice and we are able to observe results from both of the tosses. The state space from this experiment is given by $\Omega = \{HH, HT, TH, TT\}$, where H and T denote a toss resulting in a head and a tail, respectively. The information structure on Ω associated with the observation above is the collection

$$
\begin{aligned}
&\{\emptyset, \{HH\}, \{HT\}, \{TH\}, \{TT\}, \{HH, HT\}, \{HH, TH\}, \\
&\{HH, TT\}, \{HT, TH\}, \{HT, TT\}, \{TH, TT\}, \{HH, HT, TH\}, \\
&\{HH, HT, TT\}, \{HH, TH, TT\}, \{HT, TH, TT\}, \\
&\{HH, HT, TH, TT\}\},
\end{aligned}
\tag{2.1}
$$

where $\emptyset$ is the empty set that represents the event that never occurs. For example, the event $\{HT, TH\}$ represents the event that the result in the first toss is different from that in the second toss and $\{HT, TT\}$ the event that the second toss produces a tail. The probability assigned to each event is naturally defined as the ratio of the number of outcomes in the event to the total number of outcomes, which is 4.

A stock price at a future time is another example. The state space in this case is the set of all possible prices at the future time. One possible information structure is the collection of all possible prices and all possible price ranges

---

[1] An information structure is sometimes called a σ-algebra or a σ-field in probability literature.

of the stock as well. However, the probability measure will depend on the behavior of the stock and may be estimated through some historical data.

A probability space associated with an experiment is a triple $(\Omega, \mathcal{F}, P)$, where Ω is the state space, $\mathcal{F}$ is an information structure on Ω, and P is a probability measure on $\mathcal{F}$.

As pointed out earlier, an information structure $\mathcal{F}$ represents events that may be observed. Thus, it must have the following properties:

1. The empty set $\emptyset$ and the whole space Ω belong to $\mathcal{F}$. The empty set $\emptyset$ represents the event that never occurs and the whole space Ω represents the event that occurs with certainty;

2. If a finite or countable number of events $F_1, F_2, \cdots$ belong to $\mathcal{F}$, then the union of these events $\cup_{n=1}^{\infty} F_n$ belongs to $\mathcal{F}$. The union of a collection of events is the event that at least one of the events in the collection occurs. Thus, the interpretation of this property is that if each of the events $F_n, n = 1, 2, \cdots$, may occur then we may observe the event that at least one of the events $F_n, n = 1, 2, \cdots$, occurs. Further, the intersection of these events $\cap_{n=1}^{\infty} F_n = \{\omega; \omega \in F_n \text{ for all } n\}$ belongs to $\mathcal{F}$, too. The intersection $\cap_{n=1}^{\infty} F_n$ is the event that all the events $F_n, n = 1, 2, \cdots$, occur simultaneously;

3. If F is in $\mathcal{F}$, the complement $F^c = \{\omega \in \Omega; \omega \text{ does not belong to } F\}$ is in $\mathcal{F}$. The complement F^c of an event F represents the event that the event F does not occur. This property thus means that if we can observe that an event may occur, we can also observe that the same event might not occur.

The probability measure P assigns probabilities to the event from the information structure $\mathcal{F}$ and hence it must satisfy certain conditions. In particular, the probability measure P satisfies the following properties:

1. $P(\emptyset) = 0$, $P(\Omega) = 1$. For any event F in $\mathcal{F}$, $0 \leq P(F) \leq 1$. Recalling that $\emptyset$ represents the event that never occurs and Ω represents the event that occurs with certainty, the relative frequency of $\emptyset$ is 0, the relative frequency of Ω is 1, and the relative frequency of any other event will be in between. The property stated here simply reflects this perception;

2. The probability of the union of two mutually exclusive events F_1 and F_2 (i.e. events whose intersection is empty) is the sum of their probabilities. Mathematically, if $F_1 \cap F_2 = \emptyset$, then $P(F_1 \cup F_2) = P(F_1) + P(F_2)$. This property is referred to as additivity. Further, if a sequence of events $F_n. n = 1, 2, \cdots$, are mutually exclusive (i.e., no two events have an outcome in common), then

$$P(\cup_{n=1}^{\infty} F_n) = \sum_{n=1}^{\infty} P(F_n).$$

The additivity property implies that a probability measure as a function of an event is 'increasing' in the sense that for two events F_1 and F_2 with $F_1 \subseteq F_2$, $P(F_1) \leq P(F_2)$. To see this, write $F_2 = F_1 \cup (F_1^c \cap F_2)$. Then the events F_1 and $F_1^c \cap F_2$ are mutually exclusive, so that $P(F_2) = P(F_1) + P(F_1^c \cap F_2) \geq P(F_1)$.

Since an information structure is a collection of subsets of a state space, it is usually the case that more than one information structure may be associated with a given state space. That is, there is more than one way to observe the outcome of an experiment. The information we may obtain from an experiment depends on how we observe it. The same experiment may lead to different information structures. Take the coin-toss example described above. Suppose that one can see the first toss but not the second. The observable events in this situation are those in the collection

$$\{\emptyset, \{HH, HT\}, \{TH, TT\}, \{HH, HT, TH, TT\}\}, \qquad (2.2)$$

which forms an information structure that is different from that of (2.1), where one observes both tosses. As a second example, suppose that one can observe the price of a stock over a certain period of time in two different ways: one by continuously tracking the stock price and the other by recording only the closing price on each trading date. In this case, the state space is the same and may be represented by the set of all possible price paths over this time period. However, the information structure that represents continuous tracking is different from the information structure that represents discrete sampling. Further, the former contains more information than the latter. Thus, in the case where there are more than one information structure on a state space, it should be understood that each information structure is composed of all the events one could possibly observe under the observation method associated with it.

Suppose now that there are two information structures on a state space. One may wish to know whether one contains more information than the other. In what follows, we describe the situation where one information structure contains more information than another and show that sometimes neither information structure contains more information than the other. Suppose that all events in an information structure $\mathcal{F}_1$ are also events in another information structure $\mathcal{F}_2$. Notationally, we write $\mathcal{F}_1 \subseteq \mathcal{F}_2$. In this case, if an event A occurs according to $\mathcal{F}_1$, then it also occurs according to $\mathcal{F}_2$. Thus, $\mathcal{F}_2$ contains no less information than $\mathcal{F}_1$ and we say that $\mathcal{F}_2$ is *finer* than $\mathcal{F}_1$, or $\mathcal{F}_1$ is *coarser* than $\mathcal{F}_2$. Further, if there is at least one event in $\mathcal{F}_2$ that is not in $\mathcal{F}_1$, then $\mathcal{F}_2$ contains more information than $\mathcal{F}_1$ and we say that $\mathcal{F}_2$ ($\mathcal{F}_1$) is *strictly finer (coarser)* than $\mathcal{F}_1$ ($\mathcal{F}_2$). In the coin-toss example, the information structure (2.1) contains more information than the information structure (2.2), so that (2.1) is strictly finer than (2.2). Consider now two special information structures on a state space Ω. Let $\mathcal{F}_\phi = \{\emptyset, \Omega\}$. $\mathcal{F}_\phi$ which contains no information is the coarsest information structure. Let $\mathcal{F}_\Omega$

be the set of all subsets of Ω. $\mathcal{F}_\Omega$ which contains all the information from the underlying state space Ω is the finest information structure. Any other information structure on Ω is finer than $\mathcal{F}_\phi$ and coarser than $\mathcal{F}_\Omega$.

Example 2.1 Partition on a Finite State Space
Consider a state space Ω which has only a finite number of states, i.e., $\Omega = \{\omega_1, \omega_2, \cdots, \omega_k\}$. A (finite) *partition* of Ω is a collection of mutually exclusive subsets whose union covers Ω. For example, if $\Omega = \{1, 2, 3, 4\}$, then $\{\{1\}, \{2\}, \{3, 4\}\}$ is a partition.

In what follows, we show that for each information structure on the finite state space Ω there is a unique partition on Ω corresponding to that information structure and vice versa. This one-to-one correspondence allows for using a partition to represent an information structure $\mathcal{F}$, where events in the partition are the 'smallest' in the sense that no other event in $\mathcal{F}$ is contained in these events. Moreover, any other event in $\mathcal{F}$ is the union of some of these smallest events. We begin with a given information structure $\mathcal{F}$ and identify the corresponding partition. Let ω be a state in Ω. First, find the smallest event F_ω containing ω in $\mathcal{F}$, meaning that there is no other event in $\mathcal{F}$ that is contained in F_ω. The choice of F_ω is unique. Suppose that there is another 'smallest' event F'_ω containing ω. Then the intersection $F_\omega \cap F'_\omega$ is also an event in $\mathcal{F}$ (Property 2 of an information structure) and it contains ω. Since F_ω and F'_ω are not the same, $F_\omega \cap F'_\omega$ is contained in F_ω and is strictly smaller, which contradicts to the selection of F_ω. Therefore, there is only one smallest event containing ω. If $F_\omega = \Omega$, we find the partition and stop. If $F_\omega \neq \Omega$, choose a state ω' outside F_ω and find the smallest event $F_{\omega'}$ containing ω' in $\mathcal{F}$. F_ω and $F_{\omega'}$ are mutually exclusive. If not, the difference[2] $F_\omega \backslash F_{\omega'}$ is a strictly smaller event contained in F_ω, a contradiction to the selection of F_ω or $F_{\omega'}$. If F_ω and $F_{\omega'}$ cover Ω, these events form a partition and we stop. Otherwise, continue in this way until we cover every state in Ω with 'smallest' events. Since Ω has only a finite number of states, it takes only a finite number of steps to find a partition. It is easy to see that the partition obtained in the above procedure is unique. If there is another partition such that events in the partition are in $\mathcal{F}$ and there is no other event in $\mathcal{F}$ contained in these events, a non-empty intersection of an event in the former partition and an event in the latter exists. That intersection is strictly smaller than one of the larger events, a contradiction to the selection of a partition. Conversely, for a given partition, all possible unions of events in the partition form an information structure in which the events in the partition are the smallest. Therefore, for a finite state space, there is a one-to-one correspondence between its information structures and partitions. Further, it is easy to show that the implication relationship is preserved, that is, an information structure is finer than another information structure if and only if the corresponding partition of the former

---

[2] $A \backslash B$ is the event containing all the outcomes that are in A but not in B.

is finer than that of the latter. A partition is finer than another if each event in the former is a subset of an event of the latter.

As an example, consider now the state space $\Omega = \{1, 2, 3, 4\}$. The partition $\{\{1\}, \{2\}, \{3, 4\}\}$ generates the information structure

$$\{\emptyset, \{1\}, \{2\}, \{1, 2\}, \{3, 4\}, \{1, 3, 4\}, \{2, 3, 4\}, \Omega\}.$$

For a coarser partition $\{\{1, 2\}, \{3, 4\}\}$, the corresponding information structure is

$$\{\emptyset, \{1, 2\}, \{3, 4\}, \Omega\},$$

which is coarser than the previous information structure. Consider now a new partition $\{\{1\}, \{2, 3\}, \{4\}\}$. the information structure corresponding to this partition is

$$\{\emptyset, \{1\}, \{4\}, \{2, 3\}, \{1, 4\}, \{1, 2, 3\}, \{2, 3, 4\}, \Omega\},$$

which is unrelated to either of the information structures above, meaning that it is neither finer nor coarser than either information structure. $\qquad\square$

Example 2.2 Consider a stock over a two-day period with current price of 100. The stock price will move up or down 10 points on day one and will move up or down 5 points on day two from its day one price. Thus, there are two possible prices on day one: 90, 110, and four possible prices on day two: 85, 95, 105, 115. To model the stock price at the end of the period to incorporate the price information, we let $\Omega = \{85, 95, 105, 115\}$. The partition of the information structure $\mathcal{F}_2$ on day two is clearly $\{\{85\}, \{95\}, \{105\}, \{115\}\}$. Since on day one if the price is 90, we only know that the price the next day will be one of 85 or 95. Thus this event is represented by $\{85, 95\}$. Similarly, if the event when the price is 110 is represented by $\{105, 115\}$. Hence, the partition of the information structure $\mathcal{F}_1$ on day one is $\{\{85, 95\}, \{105, 115\}\}$. Obviously, the information structure $\mathcal{F}_2$ is finer than the information structure $\mathcal{F}_1$. Further, the events $\{85\}, \{95\}, \{105\}, \{115\}$ are unobservable events with respect to day one's information structure $\mathcal{F}_1$ but are observable events with respect to day two's information structure $\mathcal{F}_2$.

$\qquad\square$

When dealing with a finite state space and more than one information structure, we can express the relations among them with an information tree. A tree shows the flow of events belonging to different information structures, in the order of the coarsest to the finest. It includes nodes which represent possible events and branches which indicate an implication relation between two events belonging to two consecutive information structures. As we will see throughout this book, a tree expression is very useful when calculating conditional probabilities and conditional expectations of finite-state random variables as well as discrete-time finite-state stochastic processes that will be introduced later.

In Figure 2.1., we show how the information structures in Example 2.2 is expressed in terms of an information tree.

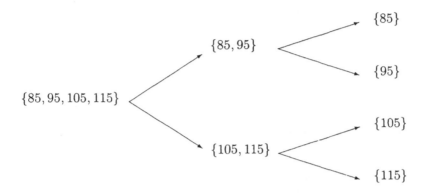

Figure 2.1. The price of a stock over a two-day period.

2.2 RANDOM VARIABLES, MOMENTS AND TRANSFORMS

In many practical situations, we are interested in some real-valued function of the outcomes of an experiment. For example, if a bet is placed on the results of the tosses in the coin-toss example described in Section 2.1, we may be interested in the dollar amount won as a result of the betting on each outcome. Other examples include stock prices, interest rates, and economic indices, whose values are contingent on the state of an economy. A real-valued function of the outcomes of an experiment is called a random variable. The precise mathematical definition is given as follows.

Suppose that $(\Omega, \mathcal{F}, P)$ is a probability space and X is a real-valued function defined on the state space Ω, i.e., for each outcome $\omega \in \Omega$, $X(\omega)$ is a real value. The function $X : \Omega \to \mathcal{R}$, where $\mathcal{R}$ is the set of real numbers, is called a random variable, if for any real number x the set $\{\omega; X(\omega) \le x\}$ is an event in $\mathcal{F}$. Sometimes we simply write $\{X \le x\} \in \mathcal{F}$. Since $\{X \le x\}$ represents the situation that the value of X is less than or equal to x, the requirement that $\{X \le x\}$ is an event in $\mathcal{F}$ is necessary[3]. This requirement implies that for any $y < x$, the events $\{X > x\}$, $\{y < X \le x\}$, $\{X = x\}$, $\{X \ge x\}$ and $\{X < x\}$ all are in $\mathcal{F}$. The derivation is quite

---

[3] This requirement is usually called $\mathcal{F}$-measurability in probability literature.

simple. $\{X > x\} \in \mathcal{F}$ follows from $\{X > x\} = \{X \le x\}^c$ and Property 3 of the definition of information structure. $\{y < X \le x\} \in \mathcal{F}$ is implied by $\{y < X \le x\} = \{X \le x\} \cap \{X > y\}$ and Property 2 of the definition of information structure. $\{X = x\} \in \mathcal{F}$ follows from $\{X = x\} = \cap_{n=1}^{\infty}\{x - 1/n < X \le x\}$ and also Property 2. $\{X \ge x\} \in \mathcal{F}$ follows immediately from $\{X \ge x\} = \{X > x\} \cup \{X = x\}$ and Property 2. Finally, $\{X < x\} \in \mathcal{F}$ follows from $\{X < x\} = \{X \ge x\}^c$ and Property 3.

It is clear from the definition that a random variable is always tied to an information structure. In the case where there are two or more information structures on a state space, a real-valued function can be a random variable with respect to one information structure but not with respect to others. However, any real-valued function is a random variable with respect to the finest information structure $\mathcal{F}_{\Omega}$ that consists of all subsets of the state space. The following example illustrates this situation.

Example 2.3 (Continuation of Example 2.2)
As in Example 2.2, a stock over a two-day period with current price of 100 has four possible prices on day two: $85, 95, 105, 115$. The state space is $\Omega = \{85, 95, 105, 115\}$. The partition of the information structure $\mathcal{F}_2$ on day two is $\{\{85\}, \{95\}, \{105\}, \{115\}\}$, while the partition of the information structure $\mathcal{F}_1$ on day one is $\{\{85, 95\}, \{105, 115\}\}$. Let X be the stock price on day two. Thus, X is a real-valued function on Ω. X is a random variable on $(\Omega, \mathcal{F}_2, P)$, since $\mathcal{F}_2$ is the collection of all subsets of Ω. However, X is not a random variable on $(\Omega, \mathcal{F}_1, P)$ since (for example) $\{X \le 105\} = \{85, 95, 105\}$ is not an event in $\mathcal{F}_1$.

Consider now a situation where a trader on day one is willing to pay 90 if the stock price on day two will be 85 or 95, and is willing to pay 110 if the price on day two will be 105 or 115. Let Y denote this price. It is easy to verify that Y is a random variable on $(\Omega, \mathcal{F}_1, P)$.

$\square$

Given a random variable X on $(\Omega, \mathcal{F}, P)$, let us now consider a specific information structure that contains all the information we are able to obtain from X. Suppose that X is the only source from which we can obtain information about an experiment. In this case, all events can be expressed in terms of a set of values of X. If two events have the same set of values of X, we can not tell that they are different. Hence, all possible events we can see through X are the subsets $\{X \le x\}$, for all real x, and events generated by unions, intersections and complementation among the subsets $\{X \le x\}$. All these events form an information structure[4]. Hence, we obtain an informa-

---

[4]Mathematically, this information structure is the coarsest or smallest information structure containing $\{X \le x\}$, for all real x.

tion structure that contains all the information we can obtain from the random variable X. We refer to this information structure as the Borel information structure or natural information structure with respect to X and denote it by $\mathcal{B}_X$. Obviously, $\mathcal{B}_X$ is coarser than $\mathcal{F}$ and contains no more information than $\mathcal{F}$. This is because by the definition of a random variable, any event in $\mathcal{B}_X$ is an event in $\mathcal{F}$. The following example shows how to obtain a natural information structure with respect to a random variable.

Example 2.4   Consider an economy which has 3 possible states at time 1, i.e., $\Omega = \{\omega_1, \omega_2, \omega_3\}$. One and only one of the states will prevail at time 1. Thus the information structure from this economy is the collection of all the subsets of Ω, the finest information structure. The corresponding partition is $\{\{\omega_1\}, \{\omega_2\}, \{\omega_3\}\}$. Let S_1 represent the price of a stock at time 1. $S_1 = 110$, if ω_1 or ω_2 prevails, and $S_1 = 90$, if ω_3 prevails. Thus, the natural information structure $\mathcal{B}_S$ with respect to the stock price S_1 is given by the partition $\{\{\omega_1, \omega_2\}, \{\omega_3\}\}$, which is coarser than but not the same as $\{\{\omega_1\}, \{\omega_2\}, \{\omega_3\}\}$.

More generally, if a random variable, say X, only takes a finite number of different values $u_1, u_2, \cdots, u_k$. Let $A_i = \{X = u_i\}$, $i = 1, 2, \cdots, k$. Then $\mathcal{B}_X$ is the collection of all possible unions of the partition $\{A_1, A_2, \cdots, A_k\}$. That is, $\mathcal{B}_X$ is the information structure corresponding to the partition $\{A_1, A_2, \cdots, A_k\}$. Note that from the choice of A_i, the value of X remains constant on each A_i. The converse is also true, i.e., if the information structure is generated by a finite partition, say $\{A_1, A_2, \cdots, A_k\}$, then the value of any random variable on this information structure is constant on each A_i. If not, we may assume that the random variable X takes at least two different values $x_1 < x_2$ on A_1. Then the event $\{X \leq x_1\} \cap A_1$ is a strictly smaller event than A_1, which contradicts the fact that A_1 is an element of the partition and hence is the smallest.

$\square$

We now discuss notions and properties related to random variables. The (cumulative) distribution function (cdf) of a random variable X is defined as $F(x) = P\{X \leq x\}$, for any real x. Thus, the distribution function $F(x)$ is nondecreasing with $F(-\infty) = 0$ and $F(\infty) = 1$. The survival function (sf) of X is defined as $s(x) = P\{X > x\}$. Obviously, $s(x) = 1 - F(x)$.

If X only takes a countable number of values, say $x_0, x_1, \cdots$, we say X is a discrete random variable and $p(x_k) = P\{X = x_k\}$, $k = 0, 1, \cdots$, is called the probability (mass) function (pf) of X. A counting random variable is a discrete random variable whose values are nonnegative integers, i.e. $x_k = k$, $k = 0, 1, \cdots$. If $F(x)$ is differentiable, we say X is a continuous random variable and define $f(x) = F'(x)$ the probability density function or simply density function (pdf) of X. Thus, $f(x)$ is a nonnegative function. It is easy to see that

$$F(x) = \int_{-\infty}^{x} f(y)dy, \text{ with } \int_{-\infty}^{\infty} f(y)dy = 1.$$

More generally, we have

$$P(A) = \int_{A} f(y)dy,$$

for any event A. Heuristically, we may write $P\{X = x\} = f(x)dx$ and view $P\{X = x\}$ as the probability $P\{x \leq X < x + dx\}$ of the infinitesimally small interval $[x, x + dx)$. This approach enables us to treat a continuous random variable somewhat like a discrete random variable, which, in many situations, is easier to handle.

The expectation of a random variable is, roughly speaking, the average of the values of the random variable. For a discrete random variable X, the expectation (also called mean) of $E(X)$ is defined as

$$E(X) = \sum_{k=0}^{\infty} x_k P\{X = x_k\} = \sum_{k=0}^{\infty} x_k p(x_k). \qquad (2.3)$$

For a continuous random variable X, the expectation $E(X)$ is defined as

$$E(X) = \int_{-\infty}^{\infty} x f(x)dx. \qquad (2.4)$$

In general, the expectation $E(X)$ of a random variable X is defined as

$$E(X) = \int_{-\infty}^{\infty} x dF(x),$$

where the integral is taken in the Steiltjes[5] sense. However, most random variables we encounter are discrete, continuous, or a combination of the two. The formulas (2.3) and (2.4) will be sufficient for expectation calculations.

A real-valued function $h(X)$ of a random variable X is also a random variable, provided that the function $h(x)$ is continuous or has a countable number of discontinuous points. The expectation of $h(X)$ can be calculated in terms of the distribution of X. If X is discrete,

$$E(h(X)) = \sum_{k=0}^{\infty} h(x_k)p(x_k). \qquad (2.5)$$

If X is continuous,

$$E(h(X)) = \int_{-\infty}^{\infty} h(x)f(x)dx. \qquad (2.6)$$

[5]The Steiltjes integral is beyond the scope of this book. Interested readers are referred to Rudin (1976), p. 122.

The expectation of $h(X)$ when $h(x)$ takes certain forms plays an important role in quantifying and measuring the distribution of X. If $h(x) = x^n$, we obtain the n-th moment, $E(X^n)$, of X. If $h(x) = (x - E(X))^2$, we obtain the variance $Var(X) = E[(X - E(X))^2]$ of X. It is easy to see the relation $Var(X) = E(X^2) - [E(X)]^2$. The square root of the variance, $\sqrt{Var(X)}$, is called the standard deviation of X. The variance of a random variable X (and hence also the standard deviation) is often used to measure the spread of probability in the distribution of X.

Let now $h(x) = e^{zx}$, for a real variable z. We obtain a function $M_X(z) = E(e^{zX})$, called the moment generating function (mgf) of X. If $h(x) = e^{-zx}$, we obtain another function $\tilde{f}_X(z) = E(e^{-zX})$, called the Laplace transform of X. Obviously, $\tilde{f}_X(z) = M_X(-z)$, and hence are interchangable. If X is a counting random variable, i.e. $X = 0, 1, \cdots$, we let $h(x) = z^x$ and define the probability generating function (pgf) of X

$$P_X(z) = E(z^X) = \sum_{k=0}^{\infty} z^k p(k). \qquad (2.7)$$

It is easy to see $M_X(z) = P_X(e^z)$ and $\tilde{f}_X(z) = P_X(e^{-z})$. It can be shown that there is a one-to-one correspondence between the distribution of a random variable and its moment generating function (or the Laplace transform or the probability generating function). For this reason, these three functions are very useful tools in identifying the distribution of some random variables. An advantage is that they are deterministic functions and hence can be manipulated using analytical methods from calculus without any probabilistic arguments. Further, it is often easier to calculate the moment generating function, the Laplace transform, or the probability generating function of a random variable, than its higher-order moments. The following formulae provide the moments of a random variable in terms of its moment generating function, Laplace transform or probability generating function.

$$E(X^n) = \left. \frac{d^n M_X(z)}{dz^n} \right|_{z=0} = (-1)^n \left. \frac{d^n \tilde{f}(z)}{dz^n} \right|_{z=0}, \qquad (2.8)$$

and

$$E(X(X-1)\cdots(X-n+1)) = \left. \frac{d^n P_X(z)}{dz^n} \right|_{z=1}. \qquad (2.9)$$

The value $E(X(X-1)\cdots(X-n+1))$ is called the n-th factorial moment of X.

However, in many cases we prefer the Laplace transform to the corresponding moment generating function since the former function exists for $z \geq 0$ while the latter may not exist for $z \neq 0$

We now present some commonly used random variables and their means, variances and Laplace transforms. We also give the probability generating function if a random variable is a counting random variable.

Example 2.5 The Binomial Distribution
Consider an experiment that has only two exclusive results: success and failure. Suppose the probability of 'success' is equal to q and thus the probability of 'failure' is $1 - q$. Let us try the experiment n times and denote the total number of successes as X. Thus X is a counting random variable and is referred to as a binomial random variable. The probability function of X is

$$p(k) = P\{X = k\} = \binom{n}{k} q^k (1 - q)^{n-k}, \quad k = 0, 1, 2, \cdots, n, \quad (2.10)$$

with parameters $0 \leq q \leq 1$ and integer n. Here

$$\binom{n}{k} = \frac{n!}{k!(n-k)!}$$

is the binomial coefficient.
It can be shown that

$$E(X) = nq, \quad Var(X) = nq(1 - q),$$

and its Laplace transform and probability generating function are

$$\begin{aligned} \tilde{f}(z) &= [qe^{-z} + (1 - q)]^n, \\ P_X(z) &= [qz + (1 - q)]^n. \end{aligned} \quad (2.11)$$

We leave the derivation of the above to interested readers.

□

Example 2.6 The Poisson Distribution
A Poisson random variable X with parameter $\lambda > 0$ is a counting random variable with probability function

$$p(k) = P\{X = k\} = e^{-\lambda} \frac{\lambda^k}{k!}, \quad k = 0, 1, 2, \cdots. \quad (2.12)$$

It is easy to verify that

$$E(X) = \lambda, \; Var(X) = \lambda.$$

The Laplace transform and probability generating function are

$$\tilde{f}(z) = e^{\lambda(e^{-z} - 1)},$$

and

$$P_X(z) = e^{\lambda(z-1)},\qquad(2.13)$$

respectively. The derivation of the above formulae can be found in any elementary probability book and is left to interested readers.

□

Example 2.7 The Normal Distribution
The normal random variable X with parameters μ and $\sigma > 0$, denoted $X \sim N(\mu, \sigma^2)$, is a continuous random variable with density

$$f(x) = \frac{1}{\sqrt{2\pi}\sigma} e^{-\frac{1}{2}(\frac{x-\mu}{\sigma})^2}, \qquad -\infty < x < \infty.$$

It is well known that its mean and variance are μ and σ^2, respectively. Its Laplace transform is

$$\tilde{f}(z) = e^{-\mu z + \frac{1}{2}\sigma^2 z^2}.\qquad(2.14)$$

and can be derived as follows:

$$
\begin{aligned}
\tilde{f}(z) &= \frac{1}{\sqrt{2\pi}\sigma} \int_{-\infty}^{\infty} e^{-zx} e^{-\frac{1}{2}(\frac{x-\mu}{\sigma})^2} dx \\
&= \frac{1}{\sqrt{2\pi}\sigma} \int_{-\infty}^{\infty} e^{-\frac{x^2 - 2(\mu - \sigma^2 z)x + \mu^2}{2\sigma^2}} dx \\
&= \frac{1}{\sqrt{2\pi}\sigma} e^{-\frac{\mu^2 - (\mu - \sigma^2 z)^2}{2\sigma^2}} \int_{-\infty}^{\infty} e^{-\frac{[x - (\mu - \sigma^2 z)]^2}{2\sigma^2}} dx \\
&= e^{-\frac{\mu^2 - (\mu - \sigma^2 z)^2}{2\sigma^2}} = e^{-\mu z + \frac{1}{2}\sigma^2 z^2}.
\end{aligned}
$$

The exponential function $Y = e^X$ of the normal random variable X is called a lognormal random variable. Obviously, the random variable Y is positive and it has the density

$$g(y) = \frac{1}{\sqrt{2\pi}\sigma y} e^{-\frac{1}{2}(\frac{\ln y - \mu}{\sigma})^2}, \qquad 0 < y < \infty.$$

The the mean and variance of Y are obtainable using the Laplace transform (2.14).

$$E(Y) = E\left(e^X\right) = \tilde{f}(-1) = e^{\mu + \frac{1}{2}\sigma^2}.$$

Similarly,

$$E\left(Y^2\right) = E\left(e^{2X}\right) = \tilde{f}(-2) = e^{2\mu + 2\sigma^2}.$$

Thus

$$Var(Y) = e^{2\mu + \sigma^2}\left(e^{\sigma^2} - 1\right).$$

However, there is no closed form expression for its Laplace transform and its moment generating function does not exist.

□

Example 2.8   The Gamma Distribution
The density function of the Gamma distribution with shape parameter $\alpha > 0$ and scale parameter $\beta > 0$ is given by

$$f(x) = \frac{\beta^\alpha x^{\alpha-1} e^{-\beta x}}{\Gamma(\alpha)}, \quad x > 0, \tag{2.15}$$

where $\Gamma(\alpha) = \int_0^\infty x^{\alpha-1} e^{-x} dx$ is the complete Gamma function. If α is an positive integer, we have $\Gamma(\alpha) = (\alpha - 1)!$. For other values of the complete Gamma function, reader may refer to Abramowitz (2002).

It can easily be shown that the Laplace transform of the Gamma distribution is given by

$$\tilde{f}(z) = \left[\frac{\beta}{\beta + x}\right]^\alpha. \tag{2.16}$$

Thus its mean and variance are α/β and α/β^2.

□

Example 2.9   The Inverse Gaussian Distribution
The density function of the inverse Gaussian distribution with shape parameter $\alpha > 0$ and scale parameter $\beta > 0$ is given by

$$f_{IG}(x) = \frac{\alpha}{\sqrt{2\pi\beta x^3}} e^{-\frac{1}{2\beta x}(\beta x - \alpha)^2}, \quad x > 0. \tag{2.17}$$

We now derive the Laplace transform of the inverse Gaussian distribution and calculate its mean and variance using (2.8). Let $\tilde{f}_{IG}(z)$ be the Laplace transform. Then,

$$
\begin{aligned}
&\tilde{f}_{IG}(z) \\
=& \int_0^\infty e^{-zx} \frac{\alpha}{\sqrt{2\pi\beta x^3}} e^{-\frac{1}{2\beta x}(\beta x - \alpha)^2} dx \\
=& \int_0^\infty \frac{\alpha}{\sqrt{2\pi\beta x^3}} e^{-\frac{1}{2\beta x}[(\beta x - \alpha)^2 + 2\beta z x^2]} dx \\
=& \int_0^\infty \frac{\alpha}{\sqrt{2\pi\beta x^3}} e^{-\frac{1}{2\beta x}[(\sqrt{\beta^2 + 2\beta z} x)^2 - 2\alpha\sqrt{\beta^2 + 2\beta z} x + \alpha^2]} \\
&\times \ e^{\frac{1}{2\beta x}[2\alpha\beta x - 2\alpha\sqrt{\beta^2 + 2\beta z} x]} dx \\
=& \ e^{\frac{\alpha}{\beta}[\beta - \sqrt{\beta^2 + 2\beta z}]} \int_0^\infty \frac{\alpha}{\sqrt{2\pi\beta x^3}} e^{-\frac{1}{2\beta x}(\sqrt{\beta^2 + 2\beta z} x - \alpha)^2} dx.
\end{aligned}
$$

Noting that the above integrand is an inverse Gaussian density with parameters α^* and β^*, where $\alpha^* = \frac{\alpha\sqrt{\beta+2z}}{\sqrt{\beta}}$ and $\beta^* = \beta + 2z$, we have

$$\tilde{f}_{IG}(z) = e^{\frac{1}{2\beta}[2\alpha\beta-2\alpha\sqrt{\beta^2+2\beta z}]} = e^{\alpha[1-\sqrt{1+2z/\beta}]}. \qquad (2.18)$$

Since

$$\tilde{f}'_{IG}(z) = -\alpha(\beta^2 + 2\beta z)^{-\frac{1}{2}}e^{\alpha[1-\sqrt{1+2z/\beta}]},$$

and

$$\tilde{f}''_{IG}(z) = \alpha^2(\beta^2+2\beta z)^{-1}e^{\alpha[1-\sqrt{1+2z/\beta}]}+\alpha\beta(\beta^2+2\beta z)^{-\frac{3}{2}}e^{\alpha[1-\sqrt{1+2z/\beta}]},$$

it follows from (2.8) that

$$E(X) = \alpha\beta^{-1}, \quad E(X^2) = \alpha^2\beta^{-2} + \alpha\beta^{-2}.$$

Hence, its mean and variance are α/β and α/β^2, respectively.

Finally in this example, we give without proof the following formula which expresses the inverse Gaussian distribution function in terms of the standard normal distribution function.

$$F_{IG}(x) = N\left(\frac{\beta x - \alpha}{\sqrt{\beta x}}\right) + e^{2\alpha}N\left(-\frac{\beta x + \alpha}{\sqrt{\beta x}}\right), \qquad (2.19)$$

where $N(x)$ is the standard normal distribution function.

$\square$

We now consider a special class of random variables. Suppose $A \in \mathcal{F}$. The indicator random variable, $\mathbf{I}_A$, associated with the event A, has value equal to 1 if A occurs and value equal to 0 if A does not occur. Mathematically, $\mathbf{I}_A(\omega) = 1$, if the outcome $\omega \in A$; $\mathbf{I}_A(\omega) = 0$, otherwise. $\mathbf{I}_A$ is indeed a random variable on $(\Omega, \mathcal{F}, P)$, since for any x the set $\{\mathbf{I}_A \leq x\}$ is either $\emptyset$, A^c or Ω, where A^c is the complement of A. It is not difficult to see

$$E(\mathbf{I}_A) = P(A). \qquad (2.20)$$

$\mathbf{I}_A$ is called the indicator of A because we know from the value of $\mathbf{I}_A$ whether the event A occurs or not. The reason that we introduce indicator random variables is because the calculation of an expectation is often easier than the calculation of a probability. With the help of (2.20) we are able to transfer a probability to an expectation. In the later chapters, we will see how this approach is used to calculate barrier hitting time distributions.

2.3 MULTIVARIATE DISTRIBUTIONS

In the preceeding section, we deal with one random variable at a time. Sometimes we are interested in more than one random variable on the same probability space, and the correlation between those random variables.

We begin with two random variables. Let X and Y be random variables on a probability space $(\Omega, \mathcal{F}, P)$. The joint distribution function of X and Y is a two-variable function $F_{X,Y}(x, y)$ and is defined as follows. For each pair of real numbers x and y,

$$F_{X,Y}(x,y) = P(X \leq x, \, Y \leq y). \tag{2.21}$$

Hence, $F_{X,Y}(x, y)$ is nondecreasing in both x and y. Since $P(X \leq x) = P(X \leq x, \, Y \leq \infty)$, the marginal distribution function of X, $F_X(x) = F_{X,Y}(x, \infty)$. Similarly, the distribution function of Y, $F_Y(y) = F_{X,Y}(\infty, y)$. Further, $F_{X,Y}(-\infty, -\infty) = 0$ and $F_{X,Y}(\infty, \infty) = 1$.

If both X and Y are discrete random variables, we define the probability mass function of X and Y as

$$p(x_j, y_k) = P(X = x_j, \, Y = y_k), \; j, k = 0, 1, \cdots, \tag{2.22}$$

where x_j and y_k are the values of X and Y, respectively. The joint density function of X and Y is defined as

$$f_{X,Y}(x,y) = \frac{\partial^2 F_{X,Y}(x,y)}{\partial x \partial y}, \tag{2.23}$$

provided the above second-order derivative exists. As in the univariate (single random variable) case, we write informally

$$P\{X = x, \, Y = y\} = f_{X,Y}(x,y) dx dy$$

and view it as the probability of the infinitesimally small box $[x, x + dx) \times [y, y + dy)$. The joint density function and the joint distribution function are related by

$$F_{X,Y}(x,y) = \int_{-\infty}^{y} \int_{-\infty}^{x} f_{X,Y}(u,v) du dv.$$

Moreover,

$$f_X(x) = \int_{-\infty}^{\infty} f_{X,Y}(x,v) dv \tag{2.24}$$

and

$$f_Y(y) = \int_{-\infty}^{\infty} f_{X,Y}(u,y) du,$$

where $f_X(x)$ and $f_Y(y)$ are the (marginal) density functions of X and Y, respectively.

Consider now a function $h(X, Y)$ of the random variables X and Y, where $h(x, y)$ is a real-valued function. When X and Y are discrete the expectation of the random variable $h(X, Y)$ is given by

$$E(h(X, Y)) = \sum_{j=0}^{\infty} \sum_{k=0}^{\infty} h(x_j, y_k) p(x_j, y_k). \qquad (2.25)$$

When X and Y are jointly continuous, i.e. the joint density of X and Y exists,

$$E(h(X, Y)) = \int_{-\infty}^{\infty} \int_{-\infty}^{\infty} h(x, y) f_{X,Y}(x, y) dx dy. \qquad (2.26)$$

The simplest case is $h(x, y) = x + y$, and we have

$$E(X + Y) = E(X) + E(Y). \qquad (2.27)$$

This can be verified from (2.25) for the discrete case or (2.26) for the continuous case. It is also true in general. A special case of $h(x, y)$ plays an important role in describing the correlation between two random variables. Let

$$h(x, y) = [x - E(X)][y - E(Y)].$$

The expectation

$$Cov(X, Y) = E\{[X - E(X)][Y - E(Y)]\} \qquad (2.28)$$

is called the covariance of X and Y. It is easy to see that

$$Cov(X, Y) = E(XY) - E(X)E(Y). \qquad (2.29)$$

The covariance is a measure for the association between X and Y. If $Cov(X, Y) > 0$, it is more likely that the two random variables move in the same direction than in an opposite direction, while $Cov(X, Y) < 0$ means that it is more likely that the two random variables move in an opposite direction than the same direction. In practice, we often use the relative value of the covariance to the standard deviations of X and Y, i.e.

$$Corr(X, Y) = \frac{Cov(X, Y)}{\sqrt{Var(X)Var(Y)}}. \qquad (2.30)$$

The quantity $Corr(X, Y)$ is called the correlation coefficient between X and Y. By [6], it can be shown that

$$-1 \leq Corr(X, Y) \leq 1.$$

[6]See Rudin (1976), p. 139.

If $Corr(X, Y) = 1$, then $X = aY + b$, with $a > 0$. Thus, X and Y move in the same direction with probability one and in this case, we say that X and Y are perfectly positively correlated. Similarly, if $Corr(X, Y) = -1$, $X = aY + b$, with $a < 0$ and X and Y move in an opposite direction with probability one. We say that X and Y are perfectly negatively correlated. If $Corr(X, Y) = 0$, we say that X and Y are uncorrelated.

The moment generating function and Laplace transform of two random variables can be defined in a way similar to the univariate case. Let $h(x, y) = e^{z_1 x + z_2 y}$, for real variables z_1 and z_2. The corresponding expectation

$$M_{X,Y}(z_1, z_2) = E\{e^{z_1 X + z_2 Y}\} \qquad (2.31)$$

is a deterministic function of z_1 and z_2, and is called the moment generating function of X and Y. Similarly, if we let $h(x, y) = e^{-z_1 x - z_2 y}$, we obtain the Laplace transform of X and Y:

$$\tilde{f}_{X,Y}(z_1, z_2) = E\{e^{-z_1 X - z_2 Y}\}. \qquad (2.32)$$

The relation

$$\tilde{f}_{X,Y}(z_1, z_2) = M_{X,Y}(-z_1, -z_2)$$

is obvious. Both (2.31) and (2.32) can be calculated using (2.25) or (2.26).

The one-to-one correspondence between the joint ditribution of X and Y and its moment generating function (or the Laplace transform) still holds. Further, we may use the moment generating function or the Laplace transform of X and Y to calculate mixed moments, which are of the form $E(X^n Y^l)$ for integers n and l. For example,

$$E(XY) = \left. \frac{\partial^2 M_{X,Y}(z_1, z_2)}{\partial z_1 \partial z_2} \right|_{z_1 = 0, z_2 = 0}.$$

The discussion above can be extended to the case of several random variables. Let $X_1, X_2, \cdots, X_m$ be m random variables on a probability space $(\Omega, \mathcal{F}, P)$. The joint distribution function $F_{X_1, \cdots, X_m}$ is defined as follows. For any real numbers $x_1, \cdots, x_m$,

$$F_{X_1, \cdots, X_m}(x_1, \cdots, x_m) = P\{X_1 \le x_1, \cdots, X_m \le x_m\}. \qquad (2.33)$$

The corresponding joint density function is given by

$$f_{X_1, \cdots, X_m}(x_1, \cdots, x_m) = \frac{\partial^m F_{X_1, \cdots, X_m}(x_1, \cdots, x_m)}{\partial x_1 \cdots \partial x_m}, \qquad (2.34)$$

if the m-th order derivative exists. The marginal density functions, i.e. the density of a group of $l < m$ random variables can be obtained by integrating the joint density in (2.34) with respect to the variables that correspond to the remaining random variables. For example, if $m = 5$, the marginal density of X_1 and X_2 is a 3-fold integral

$$f_{X_1,X_2}(x_1,x_2)$$
$$= \int_{-\infty}^{\infty}\int_{-\infty}^{\infty}\int_{-\infty}^{\infty} f_{X_1,X_2,X_3,X_4,X_5}(x_1,x_2,x_3,x_4,x_5)dx_3dx_4dx_5.$$

The moment generating function of $X_1, X_2, \cdots, X_m$ is an m-variable deterministic function given by

$$M_{X_1,\cdots,X_m}(z_1,\cdots,z_m) = E\left\{e^{z_1 X_1+\cdots+z_m X_m}\right\}, \qquad (2.35)$$

and the respective Laplace transform is

$$\tilde{f}_{X_1,\cdots,X_m}(z_1,\cdots,z_m) = M_{X_1,\cdots,X_m}(-z_1,\cdots,-z_m). \qquad (2.36)$$

The one-to-one correspondence between the distribution of $X_1,\cdots,X_m$ and its moment generating function or Laplace transform still holds.

Finally in this section we present two useful multivariate distributions.

Example 2.10 The Multinomial Distribution

Let $\{B_1, B_2, \cdots, B_m\}$ be a on a probability space $(\Omega, \mathcal{F}, P)$, i.e. the events B_i, $i = 1, 2, \cdots, m$ are mutually exclusive and the union of B_i's cover the entire state space Ω. Consider now n independent trials associated with the partition, in each of which one and only one of the events B_i's occurs. Let N_i be the number of occurences of B_i. We thus obtain m correlated random variables $N_1, N_2, \cdots, N_m$ with

$$N_1 + N_2 + \cdots + N_m = n.$$

Suppose that the probability of B_i is $P(B_i) = q_i$, $i = 1, 2, \cdots, m$. Then

$$q_1 + q_2 + \cdots + q_m = 1.$$

The random variables $N_1, N_2, \cdots, N_m$ have the following probability mass function: for any nonnegative integers $n_1, n_2, \cdots, n_m$ with

$$n_1 + n_2 + \cdots + n_m = n,$$

$$\begin{aligned} p(n_1, n_2, \cdots, n_m) &= P\{N_1 = n_1, N_2 = n_2, \cdots, N_m = n_m\} \\ &= \binom{n}{n_1, n_2, \cdots, n_m} q_1^{n_1} q_2^{n_2} \cdots q_m^{n_m}, \quad (2.37) \end{aligned}$$

where

$$\binom{n}{n_1, n_2, \cdots, n_m} = \frac{n!}{n_1! n_2! \cdots n_m!}$$

is the multinomial coefficient. This distribution is called the multinomial distribution. In particular, if $m = 3$ it is called the trinomial distribtuion.

It is worth pointing out that from the constraint $N_1 + N_2 + \cdots + N_m = n$, any $m - 1$ of the random variables will completely determine the distribution. Hence, we can leave out the last random variable N_m and simply say that $N_1, N_2, \cdots, N_{m-1}$ are multinomially distributed with parameters $n, q_1, \cdots, q_{m-1}$. The corresponding probability function is such that n_m is replaced by $n - n_1 - n_2 - \cdots - n_{m-1}$ in (2.37). It is easy to see that each N_i follows a binomial distribution with parameters n and q_i. Furthermore, any group of $l < m$ random variables from $N_1, N_2, \cdots, N_m$ is also multinomially distributed. For example, N_1, N_2 is trinomially distributed with parameters n, q_1 and q_2.

The random variables $N_1, N_2, \cdots, N_m$ are negatively correlated. In fact,

$$Cov(N_i, N_j) = -nq_iq_j, \ i \neq j. \qquad (2.38)$$

Consequently,

$$Corr(N_i, N_j) = -\sqrt{\frac{q_iq_j}{(1 - q_i)(1 - q_j)}}. \qquad (2.39)$$

It is easy to derive the moment generating function of $N_1, N_2, \cdots, N_m$:

$$M_{N_1, \cdots, N_m}(z_1, \cdots, z_m) = [q_1e^{z_1} + q_2e^{z_2} + \cdots + q_me^{z_m}]^n. \qquad (2.40)$$

$\square$

Example 2.11   The Bivariate Normal Distribution
A pair of random variables X and Y are called correlated normal random variables if their joint density is given by

$$f(x, y) = \frac{1}{2\pi\sigma_1\sigma_2\sqrt{1 - \rho^2}}e^{-\frac{1}{2(1-\rho^2)}\left[\left(\frac{x-\mu_1}{\sigma_1}\right)^2 - 2\rho\left(\frac{x-\mu_1}{\sigma_1}\right)\left(\frac{y-\mu_2}{\sigma_2}\right) + \left(\frac{y-\mu_2}{\sigma_2}\right)^2\right]},$$

$$(2.41)$$

for $-\infty < x, y < \infty$. Here $-\infty < \mu_1, \mu_2 < \infty$, $\sigma_1 > 0, \sigma_2 > 0$, and $-1 \leq \rho \leq 1$. It follows from (2.24) that $X \sim N(\mu_1, \sigma_1^2)$ and $Y \sim N(\mu_2, \sigma_2^2)$. Further, $Corr(X, Y) = \rho$. Thus, all the parameters are meaningful: the μ's and σ's are the means and standard deviations of the random variables, and ρ is the correlation coefficient. The moment generating function of X and Y will be given in Example 2.17.

$\square$

2.4   CONDITIONAL PROBABILITY AND CONDITIONAL DISTRIBUTIONS

Conditional probability and conditional expectation play a very important role in the theory of stochastic processes. Roughly stated, a stochastic process is a time-dependent sequence of random variables. Many quantities of interest involve probabilities and expectations associated with these random variables, conditioned on the information available at a given time. In this section, we introduce conditional probability and conditional distribution, and discuss their properties. Conditional expectation will be discussed in the next section.

We begin with the notion of conditional probability. Let $(\Omega, \mathcal{F}, P)$ be a probability space, and let A and B be two events in $\mathcal{F}$. The conditional probability of event A given that event B has occurred (or simply A given B) is defined as

$$P\{A \mid B\} = \frac{P\{A \cap B\}}{P\{B\}}, \qquad (2.42)$$

where $A \cap B$ is the intersection of A and B. Intuitively, the conditional probability of A given B is the relative likelihood of the occurence of the portion of A in B to that of the event B. Rewriting (2.42), we have

$$P\{A \cap B\} = P\{A \mid B\}P\{B\}. \qquad (2.43)$$

Formula (2.43) is often called the multiplication rule. It states that the probability of two events occuring at the same time can be expressed as the product of their probabilities where the events are treated as though one must occur after the other. This formula is very useful since the calculation of the probability of an event is easier than that of two events.

Two events A and B are said to be independent if

$$P\{A \cap B\} = P\{A\}P\{B\}. \qquad (2.44)$$

It follows from (2.43) that if A and B are independent, then $P\{A \mid B\} = P\{A\}$. The intuitive meaning here is that the probability of event A is not affected by the occurrence of event B.

Example 2.12 (Continuation of Example 2.2)
Let $A_1 = \{85\}, A_2 = \{95\}, A_3 = \{105\}$ and $A_4 = \{115\}$. Suppose that the probabilities of A_1, A_2, A_3 and A_4 are 0.2, 0.2, 0.4, 0.2, respectively. Also let $B_1 = \{85, 95\}$ and $B_2 = \{105, 115\}$. B_1 and B_2 represent the events corresponding to day one's price. We have $P(B_1) = 0.4$ and $P(B_2) = 0.6$. Since $A_1 \subset B_1$

$$P\{A_1 \mid B_1\} = \frac{P(A_1)}{P(B_1)} = \frac{0.2}{0.4} = 0.5.$$

Similarly, we have

$$P\{A_2 \mid B_1\} = 0.5, P\{A_3 \mid B_2\} = 0.\bar{6}, P\{A_4 \mid B_2\} = 0.\bar{3}.$$

Moreover, since $P(A_1) = P(A_2) = 0.2$, the events A_1 and B_1 are not independent and neither are A_2 and B_1.

$\square$

As we have seen in Figure 2.1, information structures on a finite state space can be expressed in terms of an information tree. Modifying an information tree by adding respective probabilities, we obtain a so-called probability tree, which not only indicates the implication relations between information structures but the conditional probability of the event at the end of a branch given the event at the beginning of the branch. We may further modify the probability tree by replacing the event at each node with a quantity of interest (probabilities, expectations, etc.) associated with the event.

Figure 2.2. shows the probability tree related to Example 2.12.

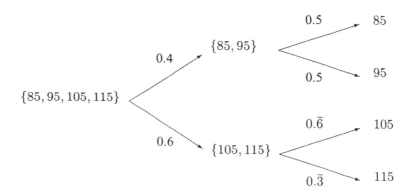

Figure 2.2.   The probability tree of the stock price over a two-day period.

☐

We now present two important identities associated with conditional probability: Bayes' Formula and the Law of Total Probability.

Bayes' Formula
For two events A and B,

$$P\{A \mid B\} = \frac{P\{B \mid A\}P\{A\}}{P\{B\}}. \qquad (2.45)$$

This identity is quite obvious from the definition of conditional probability. However, it is extremely useful in calculating many conditional probabilities when the probability of a conditional event in the reversed order is given.

The Law of Total Probability
If $\{B_1, B_2, \cdots, B_n\}$ is a partition of the probability space $(\Omega, \mathcal{F}, P)$, i.e.

B_i, $i = 1, 2, \cdots, n$ are mutually exclusive events and the union of all B_i's cover the entire state space Ω, then for any event A

$$P\{A\} = \sum_{i=1}^{n} P\{A \mid B_i\} P\{B_i\}. \tag{2.46}$$

To derive this identity we observe that

$$P\{A \mid B_i\} P\{B_i\} = P\{A \cap B_i\}, \quad A = \cup_{i=1}^{n}(A \cap B_i)$$

and

$$(A \cap B_i) \cap (A \cap B_j) = \emptyset, i \neq j.$$

By additivity, we have

$$P\{A\} = \sum_{i=1}^{n} P\{A \cap B_i\} = \sum_{i=1}^{n} P\{A \mid B_i\} P\{B_i\}.$$

The Law of Total Probability provides an algorithm to calculate unconditional probabilities by conditioning. Also, note that formula (2.46) allows $n = \infty$.

We now consider conditional distributions and related issues. Let X and Y be two random variables. The distribution of X, given that $Y = y$ (or simply $X|Y = y$) can be calculated as follows:

Assume first that both X and Y are discrete with joint probability function $p(x_j, y_k)$, $j, k = 0, 1, \cdots$. Then the conditional probability function $p_{X|Y}(x_j|y_k)$ of $X|Y = y_k$ is given by

$$p_{X|Y}(x_j|y_k) = P\{X = x_j|Y = y_k\} = \frac{P\{X = x_j, Y = y_k\}}{P\{Y = y_k\}} = \frac{p(x_j, y_k)}{p_Y(y_k)}, \tag{2.47}$$

where $p_Y(y) = \sum_{j=0}^{\infty} p(x_j, y)$ is the marginal probability function of Y. Applying Bayes' Formula (2.45), we have

$$p_{Y|X}(y_k|x_j) = \frac{p_{X|Y}(x_j|y_k)p_Y(y_k)}{p_X(x_j)}, \tag{2.48}$$

where $p_X(x) = \sum_{k=0}^{\infty} p(x, y_k)$ is the marginal probability function of X. The formula (2.48) provides the conditional probability function of $Y|X = x_j$.

Assume next that both X and Y are continuous with joint density $f_{X,Y}(x, y)$. Then the conditional density $f_{X|Y}(x|y)$ of $X|Y = y$ can be derived heuristically as follows.

$$
\begin{aligned}
f_{X|Y}(x|y)dx &= P\{X = x|Y = y\} \\
&= \frac{P\{X = x, Y = y\}}{P\{Y = y\}} = \frac{f_{X,Y}(x,y)dxdy}{f_Y(y)dy} \\
&= \frac{f_{X,Y}(x,y)}{f_Y(y)}dx.
\end{aligned}
$$

Therefore,

$$f_{X|Y}(x|y) = \frac{f_{X,Y}(x,y)}{f_Y(y)}. \tag{2.49}$$

Similar to the discrete case, we have the continuous version of Bayes' Formula

$$f_{Y|X}(y|x) = \frac{f_{X|Y}(x|y)f_Y(y)}{f_X(x)}. \tag{2.50}$$

More generally, consider a conditional probability $P\{A|Y = y\}$, where A is an event. When Y is discrete,

$$P\{A|Y = y_k\} = \frac{P\{A, Y = y_k\}}{p_Y(y_k)}, \tag{2.51}$$

and when Y is continuous,

$$P\{A|Y = y\} = \frac{\partial P\{A, Y \le y\}}{\partial y} / f_Y(y). \tag{2.52}$$

Formulae (2.51) and (2.52) provide a general algorithm in deriving the conditional distribution function $F_{X|Y}(x|y)$ of $X|Y = y$ for other situations. Let $A = \{X \le x\}$. Then, if Y is discrete the conditional distribution function

$$\begin{aligned} F_{X|Y}(x|y_k) &= \frac{P\{X \le x, Y = y_k\}}{p_Y(y_k)} \\ &= \frac{F_{X,Y}(x, y_k) - F_{X,Y}(x, y_k-)}{p_Y(y_k)}, \end{aligned}$$

where y_k- denotes the largest value y among $y_0, y_1, \cdots$, satisfying $y < y_k$. If Y is continuous,

$$F_{X|Y}(x|y) = \frac{\partial F_{X,Y}(x, y)}{\partial y} / f_Y(y).$$

Formulae (2.51) and (2.52) can also be used to calculate unconditional probabilities when conditional probabilities are given. Suppose that we want to calculate $P(A)$ for some event A. If Y is discrete it follows from (2.51) that

$$P(A) = \sum_{k=0}^{\infty} P(A, Y = y_k) = \sum_{k=0}^{\infty} P(A|Y = y_k)p_Y(y_k). \tag{2.53}$$

If Y is continuous, we use (2.52) and have

$$P(A) = \int_{-\infty}^{\infty} \frac{\partial P\{A, Y \le y\}}{\partial y} dy = \int_{-\infty}^{\infty} P\{A|Y = y\}f_Y(y)dy. \tag{2.54}$$

Formulae (2.53) and (2.54) are other versions of the Law of Total Probability (2.46).

We now extend the conditional distribution of X given a single random variable Y to the conditional distribution of X, given several ($m > 1$) other random variables. For simplicity we take $m = 3$ and assume that either all the random variables are discrete or all are continuous. In the discrete case, the conditional probability function of $X|Y = y, Z = z, W = w$ is given by

$$
\begin{aligned}
p_{X|Y,Z,W}(x|y,z,w) &= P\{X = x|Y = y, Z = z, W = w\} \\
&= \frac{P\{X = x, Y = y, Z = z, W = w\}}{P\{Y = y, Z = z, W = w\}} \\
&= \frac{p_{X,Y,Z,W}(x,y,z,w)}{p_{Y,Z,W}(y,z,w)}.
\end{aligned}
\tag{2.55}
$$

Similarly for the continuous case, the conditional density of $X|Y = y, Z = z, W = w$ is given by

$$
f_{X|Y,Z,W}(x|y,z,w) = \frac{f_{X,Y,Z,W}(x,y,z,w)}{f_{Y,Z,W}(y,z,w)}.
\tag{2.56}
$$

Two random variables X and Y are said to be independent if

$$
P\{X \le x, Y \le y\} = P\{X \le x\}P\{Y \le y\},
\tag{2.57}
$$

for all real numbers x and y. Equation (2.57) is equivalent to

$$
F_{X,Y}(x,y) = F_X(x)F_Y(y).
\tag{2.58}
$$

Thus, an equivalent definition of independence is that the joint distribution function is the product of its marginal distribution functions. Also note that (2.57) implies that the events $\{X \le x\}$ and $\{Y \le y\}$ are independent. Recalling that $\{X \le x\}$ and $\{Y \le y\}$ are the seeds that generate the natural information structures $\mathcal{B}_X$ and $\mathcal{B}_Y$, respectively, we see that any event generated from X is independent of any event generated from Y. That is, for events $A \in \mathcal{B}_X$ and $B \in \mathcal{B}_Y$,

$$
P\{A \cap B\} = P\{A\}P\{B\}.
\tag{2.59}
$$

Further, if X and Y are independent, the conditional distribution function

$$
\begin{aligned}
F_{X|Y}(x|y) &= \frac{P\{X \le x, Y = y\}}{P\{Y = y\}} = \frac{P\{X \le x\}P\{Y = y\}}{P\{Y = y\}} \\
&= P\{X \le x\} = F_X(x),
\end{aligned}
\tag{2.60}
$$

i.e., the conditional distribution function is the same as the unconditional distribution function.

If X and Y are both discrete with joint probability function $p(x_j, y_k)$, X and Y are independent if and only if

$$p(x_j, y_k) = p_X(x_j)p_Y(y_k). \tag{2.61}$$

Thus, combining with (2.47), we have

$$p_{X|Y}(x_j|y) = p_X(x_j). \tag{2.62}$$

The equivalence of (2.57) and (2.61) can be proved as follows:
If (2.57) holds, it is clear from (2.59) that

$$
\begin{aligned}
p(x_j, y_k) &= P\{X = x_j, Y = y_k\} \\
&= P\{X = x_j\}P\{Y = y_k\} = p_X(x_j)p_Y(y_k).
\end{aligned}
$$

Conversely, if (2.61) holds,

$$
\begin{aligned}
&\quad P\{X \le x, Y \le y\} \\
&= \sum_{x_j \le x, y_k \le y} p(x_j, y_k) \\
&= \sum_{x_j \le x, y_k \le y} p_X(x_j)p_Y(y_k) = \sum_{x_j \le x} p_X(x_j) \sum_{y_k \le y} p_Y(y_k) \\
&= P\{X \le x\}P\{Y \le y\}.
\end{aligned}
$$

Thus, X and Y are independent.

If X and Y are both continuous with joint density function $f_{X,Y}(x, y)$, the independence of X and Y is equivalent to

$$f_{X,Y}(x, y) = f_X(x)f_Y(y), \tag{2.63}$$

where $f_X(x)$ and $f_Y(y)$ are the density functions of X and Y, respectively. The proof is the same as the discrete case except that summations are replaced by integrations. An immediate implication of (2.63) is

$$f_{X|Y}(x|y) = f_X(x), \tag{2.64}$$

the conditional density is equivalent to the unconditional density.

The case of several random variables can be discussed similarly. Random variables $X, Y, Z, \cdots, W$ are independent if and only if

$$F_{X,Y,Z,\cdots,W}(x, y, z, \cdots, w) = F_X(x)F_Y(y)F_Z(z) \cdots F_W(w). \tag{2.65}$$

Under independence, all the conditional distributions are the same as the corresponding unconditional distributions. However, it is worth pointing out that independence of all pairs does not imply independence of all random variables. For example, three random variables X, Y, Z may be pairwise independent but not mutually independent. This can be seen in the following simple example.

Example 2.13   Let X, Y, Z be random variables on a 4-state probability space with values given in the following table.

ω_i	Prob.	X	Y	Z
ω_1	0.25	1	0	0
ω_2	0.25	0	1	0
ω_3	0.25	0	0	1
ω_4	0.25	1	1	1

It is easy to see that $F_X(0) = F_Y(0) = 0.5$ and $F_X(1) = F_Y(1) = 1$. Moreover, $F_{X,Y}(0,0) = 0.25, F_{X,Y}(0,1) = 0.5, F_{X,Y}(1,0) = 0.5$, and $F_{X,Y}(1,1) = 1$. Thus, $F_{X,Y}(x,y) = F_X(x)F_Y(y)$, for any $x, y = 0, 1$. Hence, X and Y are independent. Similarly, X and Z are independent, so are Y and Z. We now show that X, Y and Z are not independent. To see this, take the point $X = Y = Z = 0$. From the above table, they can not be zero simultaneously. Hence, $F_{X,Y,Z}(0,0,0) = 0$. On the other hand, $F_X(0)F_Y(0)F_Z(0) = 0.5 \times 0.5 \times 0.5 \neq 0$. Therefore the random variables X, Y and Z are not independent.

$\square$

We now illustrate how to derive a conditional distribution from a multivariate distribution, using the trinomial distribution and the bivariate normal distribution as examples.

Example 2.14   The Trinomial Distribution
If $m = 3$ in Example 2.10, the random variables N_1 and N_2 follow a trinomial distribution with parameters n, q_1 and q_2. Thus,

$$p(n_1, n_2) = \binom{n}{n_1, n_2, n_3} q_1^{n_1} q_2^{n_2} q_3^{n_3}, \tag{2.66}$$

where $n_3 = n - n_1 - n_2$ and $q_3 = 1 - q_1 - q_2$. Consider now the conditional distribution of $N_1 | N_2 = n_2$. It follows from (2.47) that

$$p_{N_1|N_2}(n_1|n_2) = \frac{p(n_1, n_2)}{p_{N_2}(n_2)}$$

$$= \frac{\binom{n}{n_1, n_2, n_3} q_1^{n_1} q_2^{n_2} q_3^{n_3}}{\binom{n}{n_2} q_2^{n_2}(1 - q_2)^{n - n_2}}$$

$$= \binom{n - n_2}{n_1} \left(\frac{q_1}{1 - q_2} \right)^{n_1} \left(\frac{1 - q_1 - q_2}{1 - q_2} \right)^{n - n_2 - n_1},$$

$$(2.67)$$

for $n_1 = 0, 1, \cdots, n - n_2$, where

$$\binom{n}{k} = \binom{n}{k, n - k}.$$

Hence, the conditional random variable $N_1 | N_2 = n_2$ is a binomial random variable with parameters $n - n_2$ and $q = \frac{q_1}{1 - q_2}$. It is clear that N_1 and N_2 are not independent.

□

Example 2.15   The Bivariate Normal Distribution-continued
As in Example 2.11, the density of bivariate normal random variables X and Y is given by

$$f(x, y) = \frac{1}{2\pi \sigma_1 \sigma_2 \sqrt{1 - \rho^2}} e^{-\frac{1}{2(1 - \rho^2)} \left[\left(\frac{x - \mu_1}{\sigma_1} \right)^2 - 2\rho \left(\frac{x - \mu_1}{\sigma_1} \right) \left(\frac{y - \mu_2}{\sigma_2} \right) + \left(\frac{y - \mu_2}{\sigma_2} \right)^2 \right]},$$

from which it follows that $X \sim N(\mu_1, \sigma_1^2)$ and $Y \sim N(\mu_2, \sigma_2^2)$.
   The conditional density of $X | Y = y$ can be calculated as follows:

$$f_{X|Y}(x|y) = \frac{f(x, y)}{f_Y(y)}$$

$$= \frac{\frac{1}{2\pi \sigma_1 \sigma_2 \sqrt{1 - \rho^2}} e^{-\frac{1}{2(1 - \rho^2)} \left[\left(\frac{x - \mu_1}{\sigma_1} \right)^2 - 2\rho \left(\frac{x - \mu_1}{\sigma_1} \right) \left(\frac{y - \mu_2}{\sigma_2} \right) + \left(\frac{y - \mu_2}{\sigma_2} \right)^2 \right]}}{\frac{1}{\sqrt{2\pi} \sigma_2} e^{-\frac{1}{2} \left(\frac{y - \mu_2}{\sigma_2} \right)^2}}$$

$$= \frac{1}{\sqrt{2\pi (1 - \rho^2)} \sigma_1} e^{-\frac{1}{2} \left[\frac{x - \mu_1 - \rho \frac{\sigma_1}{\sigma_2} (y - \mu_2)}{\sqrt{1 - \rho^2} \sigma_1} \right]^2}. \qquad (2.68)$$

Thus, $X | Y = y$ is a normal random variable with mean $\mu_1 + \rho \frac{\sigma_1}{\sigma_2} (y - \mu_2)$ and variance $(1 - \rho^2) \sigma_1^2$. Comparing with the marginal distribution of X, we conclude that two bivariate normal random variables X and Y are independent if and only if their correlation coefficient $\rho = 0$.

□

   We now turn to the expectation of the product of independent random variables. Let $g(X), h(Y), u(Z), \cdots, v(W)$ be the functions of the independent random variables $X, Y, Z, \cdots, W$. Then $g(X), h(Y), u(Z), \cdots, v(W)$

are also independent and

$$E\{g(X)h(Y)u(Z)\cdots v(W)\}$$
$$= E\{g(X)\}E\{h(Y)\}E\{u(Z)\}\cdots E\{v(W)\}. \qquad (2.69)$$

To see this we look at the case of three continuous random variables.

$$E\{g(X)h(Y)u(Z)\}$$
$$= \int_{-\infty}^{\infty}\int_{-\infty}^{\infty}\int_{-\infty}^{\infty} g(x)h(y)u(z)f_{X,Y,Z}(x,y,z)dxdydz$$
$$= \int_{-\infty}^{\infty}\int_{-\infty}^{\infty}\int_{-\infty}^{\infty} g(x)h(y)u(z)f_X(x)f_Y(y)f_Z(z)dxdydz$$
$$= \left(\int_{-\infty}^{\infty} g(x)f_X(x)dx\right)\left(\int_{-\infty}^{\infty} h(y)f_Y(y)dy\right)\left(\int_{-\infty}^{\infty} u(z)f_Z(z)dz\right)$$
$$= E\{g(X)\}E\{h(Y)\}E\{u(Z)\}.$$

The derivation in other cases can be done similarly and is omitted here.

The formula (2.69) has many applications. Let $g(X) = [X - E(X)]$ and $h(Y) = [Y - E(Y)]$. Then $E\{g(X)h(Y)\} = Cov(X,Y)$ (see (2.28)). Thus (2.69) implies that if X and Y are independent, then

$$Cov(X,Y) = E\{X - E(X)\}E\{Y - E(Y)\} = 0.$$

Thus if two random variables are independent, they are uncorrelated.

If $X_1, X_2, \cdots, X_m$, are independent random variables, then the sum of random variables $X = \sum_{i=1}^{m} X_i$ has

$$Var\left(\sum_{i=1}^{m} X_i\right) = \sum_{i=1}^{m} Var(X_i) + 2\sum_{i>j} Cov(X_i, X_j)$$
$$= \sum_{i=1}^{m} Var(X_i). \qquad (2.70)$$

The joint moment generating function of $X_1, X_2, \cdots, X_m$ can also be calculated easily. In fact,

$$M_{X_1,\cdots,X_m}(z_1,\cdots,z_m) = \prod_{i=1}^{m} M_{X_i}(z_i), \qquad (2.71)$$

i.e. the joint moment generating function of $X_1, X_2, \cdots, X_m$ is the product of the individual moment generating functions. To see this, we have

$$M_{X_1,\cdots,X_m}(z_1,\cdots,z_m) = E\{e^{z_1 X_1 + \cdots + z_m X_m}\}$$
$$= E\{e^{z_1 X_1}\cdots e^{z_m X_m}\}$$
$$= E\{e^{z_1 X_1}\}\cdots E\{e^{z_m X_m}\}$$
$$= M_{X_1}(z_1)\cdots M_{X_m}(z_m).$$

It can also be shown that if the joint moment generating function of $X_1, X_2,$ $\cdots, X_m$ is the product of m functions $M_1(z_1), \cdots, M_m(z_m)$, that is, the variables $z_1, \cdots, z_m$ can be separated in the joint moment generating function, then $X_1, X_2, \cdots, X_m$ are independent and $M_i(z_i)$ differs from the moment generating function of X_i up to a constant for $i = 1, \cdots, m$.

Noting that for $X = \sum_{i=1}^{m} X_i$,

$$M_X(z) = M_{X_1, \cdots, X_m}(z, z, \cdots, z),$$

we have

$$M_X(z) = \prod_{i=1}^{m} M_{X_i}(z). \tag{2.72}$$

Example 2.16   Sum of Independent Normal Random Variables

Let $X_1 \sim N(\mu_1, \sigma_1^2), X_2 \sim N(\mu_2, \sigma_2^2), \cdots, X_m \sim N(\mu_m, \sigma_m^2)$ be independent normal random variables. Example 2.7 shows that

$$M_{X_i}(z) = e^{\mu_i z + \frac{1}{2}\sigma_i^2 z^2}.$$

Thus, from (2.72) the moment generating function of the sum $X = \sum_{i=1}^{m} X_i$ is given by

$$M_X(z) = \prod_{i=1}^{m} M_{X_i}(z) = e^{\left(\sum_{i=1}^{m} \mu_i\right)z + \frac{1}{2}\left(\sum_{i=1}^{m} \sigma_i^2\right)z^2}.$$

Comparing it with the moment generating function of a normal random variable, we conclude that the sum of independent normal random variables X is also a normal random variable with mean $\sum_{i=1}^{m} \mu_i$ and variance $\sum_{i=1}^{m} \sigma_i^2$.

$\square$

2.5 CONDITIONAL EXPECTATION

In this section, we consider conditional expectation. We begin with the simplest case: the conditional expectation, $E\{X|Y = y\}$, of X, given $Y = y$. In the preceeding section, we obtained the conditional distribution of $X|Y = y$. Thus the conditional expectation is naturally defined as the expectation of X when the underlying distribution is the conditional distribution of $X|Y = y$. If X and Y are both discrete, from (2.47)

$$E\{X|Y = y\} = \sum_{j=0}^{\infty} x_j p(x_j|y) = \frac{\sum_{j=0}^{\infty} x_j p(x_j, y)}{\sum_{j=0}^{\infty} p(x_j, y)}. \tag{2.73}$$

If X and Y are both continuous, from (2.49)

$$E\{X|Y = y\} = \int_{-\infty}^{\infty} x f_{X|Y}(x|y)dx = \frac{\int_{-\infty}^{\infty} x f_{X,Y}(x, y)dx}{\int_{-\infty}^{\infty} f_{X,Y}(x, y)dx}. \tag{2.74}$$

The same idea can be extended to the conditional expectation of a function $h(X, Y)$ of X and Y, given $Y = y$. In this case,

$$E\{h(X, Y)|Y = y\} = E\{h(X, y)|Y = y\}. \qquad (2.75)$$

In other words, we treat the random variable Y in the function as a constant and take expectation with respect to the conditional distribution of $X|Y = y$.

Consider now the trinomial distribution discussed in Example 2.14. Since $N_1|N_2 = n_2$ is binomially distributed with parameters $n - n_2$ and $q = \frac{q_1}{1-q_2}$, we have

$$E\{N_1|N_2 = n_2\} = (n - n_2) \left(\frac{q_1}{1 - q_2} \right).$$

For correlated normal random variables X and Y with means μ_1, μ_2, variances σ_1^2, σ_2^2, and correlation coefficient ρ (Example 2.15),

$$E\{X|Y = y\} = \mu_1 + \rho \frac{\sigma_1}{\sigma_2}(y - \mu_2).$$

The conditional expectation $E\{X|Y = y, Z = z, \cdots, W = w\}$ can be calculated in a similar manner. To calculate it, we take expectation of X with respect to the underlying conditional distribution of $X|Y = y, Z = z, \cdots, W = w$.

It is evident that the conditional expectation $E\{X|Y = y\}$ (or $E\{h(X, Y)|Y = y\}$) is a function of y. Thus if we replace y by the random variable Y, we obtain another random variable. This random variable is called the conditional expectation of X (or $h(X, Y)$) with respect to Y and is written as $E\{X|Y\}$ (or $E\{h(X, Y)|Y\}$). Similarly, $E\{X|Y = y, Z = z, \cdots, W = w\}$ is a function of $y, z, \cdots, w$, and so when we replace $y, z, \cdots, w$ by $Y, Z, \cdots, W$, we obtain a random variable that is a function of $Y, Z, \cdots, W$. We call this random variable the conditional expectation of X with respect to $Y, Z, \cdots, W$.

In the trinomial example, we have

$$E\{N_1|N_2\} = (n - N_2) \left(\frac{q_1}{1 - q_2} \right).$$

Thus $E\{N_1|N_2\}$ is a scaled binomial random variable with parameters n and $1 - q_2$ (i.e. $\frac{1-q_2}{q_1} E\{N_1|N_2\}$ is binomial with parameters n and $1 - q_2$.). And in the bivariate normal example,

$$E\{X|Y\} = \mu_1 + \rho \frac{\sigma_1}{\sigma_2}(Y - \mu_2).$$

Hence, $E\{X|Y\}$ is again a normal random variable with mean μ_1 and variance $\rho^2 \sigma_1^2$.

Recall from Section 2.4 that we may calculate probabilities by conditioning, using the Law of Total Probability (2.46), (2.53) or (2.54). We can also calculate expectations by conditioning. The formula used for this is called the Law of Iterated Expectation or Tower Property.

The Law of Iterated Expectation (Tower Property)

For random variables X and Y,

$$E\{h(X,Y)\} = E\{E[h(X,Y)|Y]\}. \tag{2.76}$$

The inside expectation is a conditional expectation with respect to Y. To show (2.76), we assume that X and Y are continuous. Other cases can be discussed similarly.

$$
\begin{aligned}
E\{h(X,Y)\} &= \int_{-\infty}^{\infty} \int_{-\infty}^{\infty} h(x,y) f_{X,Y}(x,y) dx dy \\
&= \int_{-\infty}^{\infty} \int_{-\infty}^{\infty} h(x,y) f_{X|Y}(x|y) f_Y(y) dx dy \\
&= \int_{-\infty}^{\infty} \left(\int_{-\infty}^{\infty} h(x,y) f_{X|Y}(x|y) dx \right) f_Y(y) dy \\
&= \int_{-\infty}^{\infty} E\{h(X,y)|Y = y\} f_Y(y) dy \\
&= E\{E[h(X,Y)|Y]\}.
\end{aligned}
$$

Formula (2.76) can be extended to the case of several random variables. For instance, to calculate $E\{h(X,Y,Z)\}$ we apply the Law of Iterated Expectations twice as follows. First we apply (2.76) by conditioning on Z. Thus

$$E\{h(X,Y,Z)\} = E\{E\{h(X,Y,Z) \mid Z\}\}.$$

For the inside expectation $E\{h(X,Y,Z) \mid Z\}$, we apply (2.76) by conditioning on Y. As a result, we have

$$E\{h(X,Y,Z) \mid Z\} = E\{E[h(X,Y,Z) \mid Y,Z] \mid Z\}.$$

Therefore, we obtain

$$E\{h(X,Y,Z)\} = E\{E\{E[h(X,Y,Z) \mid Y,Z] \mid Z\}\}. \tag{2.77}$$

The formula (2.77) provides a step-by-step calculation of an unconditional expectation involving more than one random variable. First, take the conditional expectation of $h(X,Y,Z)$, given Y and Z. At this step, Y and Z are treated as constants. The resulting expectation $E[h(X,Y,Z) \mid Y,Z]$ is a function of Y and Z. Then take the conditional expectation of $E[h(X,Y,Z)|Y,Z]$, given Z, which is a function of Z. Finally, take the expectation of $E\{E[h(X,Y,Z) \mid Y,Z] \mid Z\}$ to obtain the unconditional expectation of $h(X,Y,Z)$.

We now derive the joint moment generating function of bivariate normal random variables, using the Law of Iterated Expectation.

Example 2.17 The Moment Generating Function of the Bivariate Normal Distribution

Let X and Y be bivariate normal random variables with respective means μ_1, μ_2 and variances σ_1^2, σ_2^2, and correlation coefficient ρ. It follows from Example 2.11 and Example 2.15 that $Y \sim N(\mu_2, \sigma_2^2)$ and $X|Y = y \sim N(\mu_y, \sigma_y^2)$, where $\mu_y = \mu_1 + \rho\frac{\sigma_1}{\sigma_2}(y - \mu_2)$ and $\sigma_y^2 = (1 - \rho^2)\sigma_1^2$.

Let $h(x, y) = e^{z_1 x + z_2 y}$. Then $M_{X,Y}(z_1, z_2) = E\{h(X, Y)\}$ is the joint moment generating function of X and Y.

Applying (2.76), we have

$$
\begin{aligned}
M_{X,Y}(z_1, z_2) &= E\{e^{z_1 X + z_2 Y}\} \\
&= E\{E\{e^{z_1 X + z_2 Y}|Y\}\} = E\{E\{e^{z_1 X}|Y\}e^{z_2 Y}\} \\
&= E\{M_{X|Y}(z_1)e^{z_2 Y}\} = E\{e^{\mu_Y z_1 + \frac{1}{2}\sigma_Y^2 z_1^2}e^{z_2 Y}\} \\
&= exp\left\{\mu_1 z_1 - \rho\frac{\sigma_1}{\sigma_2}\mu_2 z_1 + \frac{1}{2}(1 - \rho^2)\sigma_1^2 z_1^2\right\} E\{e^{\rho\frac{\sigma_1}{\sigma_2}z_1 Y + z_2 Y}\} \\
&= exp\left\{\mu_1 z_1 - \rho\frac{\sigma_1}{\sigma_2}\mu_2 z_1 + \frac{1}{2}(1 - \rho^2)\sigma_1^2 z_1^2\right\} M_Y\left(\rho\frac{\sigma_1}{\sigma_2}z_1 + z_2\right) \\
&= exp\left\{\mu_1 z_1 - \rho\frac{\sigma_1}{\sigma_2}\mu_2 z_1 + \frac{1}{2}(1 - \rho^2)\sigma_1^2 z_1^2\right\} \\
&\times exp\left\{\mu_2\left(\rho\frac{\sigma_1}{\sigma_2}z_1 + z_2\right) + \frac{1}{2}\sigma_2^2\left(\rho\frac{\sigma_1}{\sigma_2}z_1 + z_2\right)^2\right\} \\
&= exp\left\{\mu_1 z_1 + \mu_2 z_2 + \frac{1}{2}\left(\sigma_1^2 z_1^2 + 2\rho\sigma_1\sigma_2 z_1 z_2 + \sigma_2^2 z_2^2\right)\right\}.
\end{aligned}
$$

Thus we obtain the moment generating function of the bivariate normal distribution:

$$
M_{X,Y}(z_1, z_2) = exp\left\{\mu_1 z_1 + \mu_2 z_2 + \frac{1}{2}(\sigma_1^2 z_1^2 + 2\rho\sigma_1\sigma_2 z_1 z_2 + \sigma_2^2 z_2^2)\right\}.
$$

(2.78)

□

There is a close connection between the Law of Total Probability and the Law of Interated Expectation. In fact, we are able to use the Law of Interated Expectation to derive the Law of Total Probability. For any event A, let $\mathbf{I}_A$ be the indicator random variable of A (see Section 2.2). Recall that $P(A) = E(\mathbf{I}_A)$. By the Law of Interated Expectation, we have

$$
P(A) = E(\mathbf{I}_A) = E\{E(\mathbf{I}_A \mid Y)\}
$$

$$= \sum_{k=0}^{\infty} E(\mathbf{I}_A \mid Y = y_k) p_Y(y_k)$$

$$= \sum_{k=0}^{\infty} P(A \mid Y = y_k) p_Y(y_k),$$

if Y is discrete, and

$$
\begin{aligned}
P(A) &= E(\mathbf{I}_A) = E\{E(\mathbf{I}_A \mid Y)\} \\
&= \int_{-\infty}^{\infty} E(\mathbf{I}_A \mid Y = y) f_Y(y) dy \\
&= \int_{-\infty}^{\infty} P(A \mid Y = y) f_Y(y) dy,
\end{aligned}
$$

if Y is continuous. We have rederived (2.53) and (2.54).

We now turn to *conditional expectation given an event*. Suppose B is an event, i.e. $B \in \mathcal{F}$. Let $\mathbf{I}_B$ be the indicator random variable of B. Then, $B = \{\mathbf{I}_B = 1\}$. The conditional expectation of X given B is given by

$$E\{X \mid B\} = E\{X \mid \mathbf{I}_B = 1\}. \tag{2.79}$$

In other words, it is the expectation of X with respect to the conditional distribution of $X|B$.

If X is discrete,

$$E\{X \mid B\} = \sum_{j=0}^{\infty} x_j P\{X = x_j, B\}/P\{B\}; \tag{2.80}$$

and if X is continuous,

$$E\{X \mid B\} = \int_{-\infty}^{\infty} x f_{X|B}(x) dx, \tag{2.81}$$

where

$$f_{X|B}(x) = \frac{dP\{X \le x, B\}}{dx}/P\{B\}. \tag{2.82}$$

It is clear from (2.80) and (2.81) that in the case $B = \Omega$ we have

$$E\{X \mid \Omega\} = E(X), \tag{2.83}$$

i.e. the conditional expectation with respect to the entire state space is the same as the unconditional expectation.

We now derive the discrete version of the Law of Iterated Expectation, which involves conditional expectations with respect to events. Let $\{B_1, B_2, \cdots, B_n\}$ be a partition of the state space Ω. Introduce a random variable Y

such that $Y(\omega) = i$, if $\omega \in B_i, i = 1, \cdots, n$. Then, $P\{Y = i\} = P(B_i)$. By the Law of Iterated Expectation,

$$
\begin{aligned}
E(X) &= E\{E(X \mid Y)\} \\
&= \sum_{i=1}^{n} E(X \mid Y = i)P\{Y = i\} = \sum_{i=1}^{n} E(X \mid B_i)P(B_i) \quad (2.84)
\end{aligned}
$$

Formula (2.84) provides an algorithm to calculate expectation by conditioning on a partition. It is very useful when dealing with a finite-state probability space as we see in the following example.

Example 2.18 (Continuation of Example 2.12)
Let X be the stock price on day two, i.e., $X = 85, 95, 105, 115$ with probabilities 0.2, 0.2, 0.4, 0.2. Recall that $B_1 = \{85, 95\}$ and $B_2 = \{105, 115\}$. We have

$$ E(X \mid B_1) = 85P\{X = 85 \mid B_1\} + 95P\{X = 95 \mid B_1\} = 90, $$
$$ E(X \mid B_2) = 105P\{X = 105 \mid B_2\} + 115P\{X = 115 \mid B_2\} = 108.\bar{3}. $$

The conditional probabilities given B_1 and B_2 are expressed by the subtrees in the upper right corner and lower right corner of Figure 2.3.

Since B_1 and B_2 form a partition of $(\Omega, \mathcal{F}_1, P)$, The (unconditional) expectation of X is given by

$$ E(X) = E(X \mid B_1)P\{B_1\} + E(X \mid B_2)P\{B_2\} = 101, $$

by the Law of Iterated Expectation (2.84).

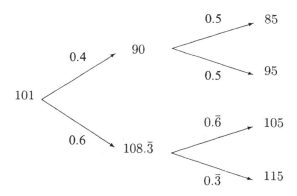

Figure 2.3. The expectation tree of the stock price over a two-day period.

Finally, we deal with a conceptually difficult notion: conditional expectation with respect to an information structure. In order to understand this notion, it is helpful for us to look again at the conditional expectation $E(X \mid Y)$ of the random variable X with respect to the random variable Y, introduced earlier in this section. Recall that $E(X \mid Y)$ is a random variable whose value is completely determined by the value of Y. In other words, this is an expectation that is based on the information from Y. Recall also that the information from Y is represented by its natural information structure $\mathcal{B}_Y$. Thus, we may view $E(X|Y)$ as the conditional expectation of X with respect to the natural information structure $\mathcal{B}_Y$, i.e.,

$$E\{X \mid \mathcal{B}_Y\} = E(X \mid Y). \tag{2.85}$$

Since Y is a random variable on the probability space $(\Omega, \mathcal{B}_Y, P)$, so is $E\{X \mid \mathcal{B}_Y\}$.

We now extend the above case to a general information structure. Let $(\Omega, \mathcal{F}, P)$ be a probability space and X a random variable on it. Also, let $\mathcal{G}$ be an information structure coarser than $\mathcal{F}$, i.e., $\mathcal{G} \subseteq \mathcal{F}$. Then $(\Omega, \mathcal{G}, P)$ is also a probability space. However, X itself may not be a random variable on $(\Omega, \mathcal{G}, P)$ since $\mathcal{G}$ is coarser than $\mathcal{F}$. The conditional expectation of X with respect to $\mathcal{G}$ is, roughly speaking, the best guess for X given the information from $\mathcal{G}$. Mathematically, the conditional expectation $E\{X \mid \mathcal{G}\}$ of X with respect to $\mathcal{G}$ is a random variable on $(\Omega, \mathcal{G}, P)$, satisfying

$$E\{E\{X \mid \mathcal{G}\} \mid G\} = E\{X \mid G\}, \tag{2.86}$$

for any event $G \in \mathcal{G}$. We need to point out that the conditional expectation of X with respect to $\mathcal{G}$, if it exists, is unique. In other words, if there is another random variable Y such that a) Y is a random variable on $(\Omega, \mathcal{G}, P)$, and b) the identity $E\{Y \mid G\} = E\{X \mid G\}$ holds for all $G \in \mathcal{G}$, then $Y = E\{X \mid \mathcal{G}\}$. The proof of the uniqueness can be found in most probability books and hence is omitted in this book.

A very useful identity can be obtained from (2.86). Let $G = \Omega$, it follows from (2.83) that

$$E\{E\{X \mid \mathcal{G}\}\} = E\{X \mid \Omega\} = E\{X\}. \tag{2.87}$$

Before we discuss the intuition of the conditional expectation with respect to an information structure, let us examine some special cases that suggest how to obtain such a conditional expectation.
(i) $\mathcal{G} = \mathcal{F}$. In this case,

$$E(X \mid \mathcal{F}) = X, \tag{2.88}$$

i.e., the conditional expectation of X with respect to the original information structure $\mathcal{F}$ is X itself. This is because X itself is a random variable on $(\Omega, \mathcal{F}, P)$ and both sides of (2.86) are exactly the same.

Property (i) basically says that we can treat a random variable as a constant when we take its conditional expectation with respect to its own information structure. This is useful in what follows. Let X be a random variable on $(\Omega, \mathcal{F}, P)$ and $\mathcal{G}$ an information structure coarser that $\mathcal{F}$. Then $E(X \mid \mathcal{G})$ is a random variable on $(\Omega, \mathcal{G}, P)$. Thus, from (2.88)

$$E\{E(X \mid \mathcal{G}) \mid \mathcal{G}\} = E(X \mid \mathcal{G}). \qquad (2.89)$$

A closely related identity is

$$E\{ZX \mid \mathcal{G}\} = ZE(X \mid \mathcal{G}), \qquad (2.90)$$

where Z is a random variable on $(\Omega, \mathcal{G}, P)$. (2.90) indicates that when taking a conditional expectation with respect to an information structure, any random variable with respect to this information structure is treated as a constant and can be taken out. The proof of (2.90) involves some results in measure theory and hence is omitted here.

(ii) $\mathcal{G} = \{\phi, \Omega\}$, i.e. no information is given. As discussed in Example 2.4 (the partition has only one element in this case), a random variable on $(\Omega, \mathcal{G}, P)$ is a constant. Thus $E(X \mid \mathcal{G})$ is a constant in this case. To satisfy (2.86), we must have $E(X \mid \mathcal{G}) = E(X)$. By conditioning X with respect to $\mathcal{G}$ we actually obtain the unconditional expectation of X.

(iii) $\mathcal{G}$ is generated by a finite partition $\{B_1, B_2, \cdots, B_n\}$. Recall that any random variable on this information structure must be constant on each B_i (Example 2.4). It follows from (2.86) that on each B_i,

$$E(X \mid \mathcal{G}) = E\{E(X \mid \mathcal{G}) \mid B_i\} = E(X \mid B_i). \qquad (2.91)$$

The first equality holds because $E(X \mid \mathcal{G})$ is constant on B_i and the second equality holds due to (2.86). Thus the value of $E(X \mid \mathcal{G})$ on each B_i is equal to the conditional expectation of X given B_i. For example, in Example 2.18, we have that $E(X \mid \mathcal{F}_2) = X$ and $E(X \mid \mathcal{F}_1)$ has two values: on B_1, $E(X \mid \mathcal{F}_1) = E(X \mid B_1) = 90$; on B_2, $E(X \mid \mathcal{F}_1) = E(X \mid B_2) = 108.\overline{3}$.

We now discuss the intuition of the conditional expectation of a random variable with respect to an information structure. Roughly speaking, if we are provided an information structure that lacks sufficient information to exactly identify the value of a random variable. The conditional expectation of the random variable with respect to the given information structure is the best estimate of the random variable, where the term best estimate is used in the sense that the corresponding estimator minimizes the mean squared error. More precisely, let X be a random variable on $(\Omega, \mathcal{F}, P)$. Thus $\mathcal{F}$ represents all the information we need to identify the value of X. In other words, given the information from $\mathcal{F}$ we know the exact value of X. Suppose now we are provided with less information which is represented by $\mathcal{G}$. In this case, we are not able to know the exact value of X. Let H be an estimate of X, given the information $\mathcal{G}$. Then, H is a random variable on $(\Omega, \mathcal{G}, P)$. Consider the least squares estimate of X given the information $\mathcal{G}$, i.e. the estimate which is the optimal solution of

$$\min_{H} E\{[X - H]^2\},$$

for all random variables H on $(\Omega, \mathcal{G}, P)$. Rewrite the above objective function as

$$
\begin{aligned}
E\{[X - H]^2\} &= E\{[X - E(X \mid \mathcal{G}) + E(X \mid \mathcal{G}) - H]^2\} \\
&= E\{[X - E(X \mid \mathcal{G})]^2\} + 2E\{[X - E(X \mid \mathcal{G})][E(X \mid \mathcal{G}) - H]\} \\
&+ E\{[E(X \mid \mathcal{G}) - H]^2\}.
\end{aligned}
$$

We have

$$
\begin{aligned}
&E\{[X - E(X \mid \mathcal{G})][E(X \mid \mathcal{G}) - H]\} \\
&= E\{E\{[X - E(X \mid \mathcal{G})][E(X \mid \mathcal{G}) - H]\} \mid \mathcal{G}\} \\
&= E\{E\{[X - E(X \mid \mathcal{G})] \mid \mathcal{G}\}[E(X \mid \mathcal{G}) - H]\} \\
&= E\{0[E(X \mid \mathcal{G}) - H]\} = 0.
\end{aligned}
$$

The first equality follows from (2.87) and the second from (2.90) where $Z = E(X \mid \mathcal{G}) - H$. The third equality follows from (2.89) since we have

$$
\begin{aligned}
E\{[X - E(X \mid \mathcal{G})] \mid \mathcal{G}\} &= E(X \mid \mathcal{G}) - E\{E(X \mid \mathcal{G}) \mid \mathcal{G}\} \\
&= E(X \mid \mathcal{G}) - E(X \mid \mathcal{G}) = 0.
\end{aligned}
$$

Thus

$$
\begin{aligned}
E\{[X - H]^2\} &= E\{[X - E(X \mid \mathcal{G})]^2\} + E\{[E(X \mid \mathcal{G}) - H]^2\} \\
&\geq E\{[X - E(X \mid \mathcal{G})]^2\},
\end{aligned}
$$

for any random variable H on $(\Omega, \mathcal{G}, P)$. Therefore, $E(X \mid \mathcal{G})$ is the least squares estimate of X. In other words, $E(X \mid \mathcal{G})$ is the best estimate for X under the expected squared error.

In the following, we discuss the Law of Iterated Expectation. Analogous to (2.76), we have the Law of Iterated Expectation in terms of two information structures.

The Law of Iterated Expectation-Information Structure

For two information structures $\mathcal{F}_1$ and $\mathcal{F}_2$, if $\mathcal{F}_2$ is finer than $\mathcal{F}_1$, then

$$E\{E(X \mid \mathcal{F}_2) \mid \mathcal{F}_1\} = E(X \mid \mathcal{F}_1). \tag{2.92}$$

The inside expectation $E(X \mid \mathcal{F}_2)$, as defined in (2.86), is a random variable on $\mathcal{F}_2$. The formula (2.92) says that to compute a conditional or unconditional expectation, we may first take the expectation by conditioning upon an appropriate finer information structure and then take the conditional expectation of that conditional expectation.

To prove identity (2.92), we observe that $E\{E(X \mid \mathcal{F}_2) \mid \mathcal{F}_1\}$ is a random variable on the information structure $\mathcal{F}_1$. Choose any event $G \in \mathcal{F}_1$. Then by (2.86), the expectation of $E\{E(X \mid \mathcal{F}_2) \mid \mathcal{F}_1\}$, given G, is

$$E\{E\{E(X \mid \mathcal{F}_2) \mid \mathcal{F}_1\} \mid G\} = E\{E(X \mid \mathcal{F}_2) \mid G\} = E(X \mid G).$$

The first equality holds because $E\{E(X \mid \mathcal{F}_2) \mid \mathcal{F}_1\}$ is the conditional expectation of $E(X \mid \mathcal{F}_2)$ and $\mathcal{F}_1 \subset \mathcal{F}_2$. The second equality holds because $E(X \mid \mathcal{F}_2)$ is the conditional expectation of X, and $G \in \mathcal{F}_2$ since $\mathcal{F}_1 \subset \mathcal{F}_2$. Thus we have
(i) $E\{E(X \mid \mathcal{F}_2) \mid \mathcal{F}_1\}$ is a random variable on $\mathcal{F}_1$;
(ii) $E\{E\{E(X \mid \mathcal{F}_2) \mid \mathcal{F}_1\} \mid G\} = E(X \mid G)$.
By the definition of the conditional expectation with respect to an information structure,

$$E\{E(X \mid \mathcal{F}_2) \mid \mathcal{F}_1\} = E(X \mid \mathcal{F}_1).$$

2.6 THE CENTRAL LIMIT THEOREM

The Central Limit Theorem is one of the greatest results in probability theory. Its applications can be found in almost every aspect of probability theory and statistics.

The Central Limit Theorem

Let X_n be a sequence of independent identically distributed (iid) random variables with common mean μ and common variance σ^2. Let

$$Y_n = \frac{X_1 + X_2 + \cdots + X_n}{n}$$

be the arithmetic average of $X_1, \cdots, X_n$. Then, for any a and b,

$$\lim_{n \to \infty} P\left\{ a < \frac{Y_n - E(Y_n)}{\sqrt{Var(Y_n)}} \leq b \right\} = \frac{1}{\sqrt{2\pi}} \int_a^b e^{-\frac{x^2}{2}} dx. \qquad (2.93)$$

The formula (2.93) implies that the limiting distribution of the standardized random variable $\frac{Y_n - E(Y_n)}{\sqrt{Var(Y_n)}}$ is normal with mean 0 and variance 1. Note that $E(Y_n) = \mu$ and $Var(Y_n) = \sigma^2/n$. Thus, for large n, the random variable Y_n is approximately normal with mean μ and variance σ^2/n. Hence, this formula provides a tool for computing probabilities associated with the arithmetic average from a large sample. It also provides a link between discrete-time stochastic processes and continuous-time stochastic processes as seen in Chapter 4.

CHAPTER 3

DISCRETE-TIME STOCHASTIC PROCESSES

3.1 STOCHASTIC PROCESSES AND INFORMATION STRUCTURES

Let $(\Omega, \mathcal{F}, P)$ be a probability space. Recall that $\mathcal{F}$ is a collection of all possible events and represents all the information contained in the probability space. Imagine that a series of experiments is performed at times $t = t_0, t_1, t_2, \cdots$. For simplicity we assume in this chapter that $t_i = i$, i.e., $t = 0, 1, 2, \cdots$, unless otherwise specified. Let $\mathcal{F}_t$ be the collection of all possible events in $\mathcal{F}$ that may occur before or at time t. Thus $\mathcal{F}_t$ represents the information up to time t. Obviously,

(i) $\mathcal{F}_t$ is an information structure coarser than $\mathcal{F}$, i.e., $\mathcal{F}_t \subseteq \mathcal{F}$, since it contains no more information than $\mathcal{F}$;

(ii) If $s < t$, then $\mathcal{F}_s$ contains no more information than $\mathcal{F}_t$, reflecting the fact that more information becomes available as time passes. Probabilitically speaking, $\mathcal{F}_s$ is coarser than $\mathcal{F}_t$, i.e. $\mathcal{F}_s \subseteq \mathcal{F}_t$. Thus, $\{\mathcal{F}_t,\ t = 0, 1, \cdots, \}$

is 'increasing'[1] in t and forms a (time-dependent) information structure or a filtration on $(\Omega, \mathcal{F}, P)$, with $\mathcal{F}_t$ representing the information up to time t.

A time-dependent information structure which satisfies (i) and (ii) is often called a *filtration* on $(\Omega, \mathcal{F}, P)$ and the quadruple $(\Omega, \mathcal{F}, \mathcal{F}_t, P)$ is called a filtered space.

An (adapted) discrete-time stochastic process on the filtered space $(\Omega, \mathcal{F}, \mathcal{F}_t, P)$ is a parametrized sequence of random variables $X(0), X(1), \cdots, X(t), \cdots$, (in short $\{X(t)\}$) such that for each fixed t, $X(t)$ is a random variable on the probability space $(\Omega, \mathcal{F}_t, P)$. Note that the information structure $\mathcal{F}$ is replaced by the information structure at time t, meaning that the value of $X(t)$ depends only on the information structure up to time t.

Example 3.1 (Continuation of Examples 2.2 and 2.3)
Define $\mathcal{F}_0 = \{\phi, \Omega\}$, the coarsest information structure. Then $\mathcal{F}_0, \mathcal{F}_1, \mathcal{F}_2$ form a time-dependent information structure. Further let $X(0) = 100$, $X(1) = 90$ if B_1 occurs and $X(1) = 110$ if B_2 occurs, and let $X(2)$ be the stock price on day two, i.e., $X(2) \in \{85, 95, 105, 115\}$. $\{X(0), X(1), X(2)\}$ is a stochastic process on $(\Omega, \mathcal{F}, \mathcal{F}_t, P)$, representing the stock price at times 0, 1 and 2.

$\square$

We now consider how information is obtained from a given stochastic process or more generally from a sequence of time-dependent random variables. Let $\{X(t), \ t = 0, 1, \cdots, \}$ be a sequence of time-dependent random variables on $(\Omega, \mathcal{F}, P)$. The information up to time t from the sequence of random variables is the information obtained from the random variables $\{X(s), \ s = 0, 1, \cdots, t\}$. As we have discussed in Chapter 2, the natural information structure $\mathcal{B}_{X(t)}$ represents the information we can obtain from the random variable $X(t)$. Thus, the information structure at time t, associated with $\{X(t), \ t = 0, 1, \cdots, \}$ is such that it is generated by all the events in $\mathcal{B}_{X(0)}, \mathcal{B}_{X(1)}, \cdots, \mathcal{B}_{X(t)}$. In other words, it is the smallest information structure containing $\mathcal{B}_{X(0)}, \mathcal{B}_{X(1)}, \cdots, \mathcal{B}_{X(t)}$. We denote this information structure as $\mathcal{B}_t$. It is easy to see that the sequence $\mathcal{B}_t, t = 0, 1, \cdots$, is increasing in t, i.e.,

$$\mathcal{B}_0 \subseteq \mathcal{B}_1 \subseteq \cdots \subseteq \mathcal{B}_t \subseteq \cdots \subseteq \mathcal{F}.$$

Thus, $\mathcal{B}_t, \ t = 0, 1, \cdots$, forms a time-dependent information structure or a filtration on $(\Omega, \mathcal{F}, P)$, called the natural information structure or the natural filtration[2] with respect to $X(t), \ t = 0, 1, \cdots$. Thus, if a sequence of random variables $X(t), \ t = 0, 1, \cdots$, defined on $(\Omega, \mathcal{F}, P)$ is given without mention of an underlying information structure, we may directly de-

---

[1]$\{\mathcal{F}_t, \ t = 0, 1, \cdots, \}$ is not necessarily strictly increasing, i.e., it is possible that $\mathcal{F}_s = \mathcal{F}_t$, for some $s < t$.
[2]it is also called the Borel filtration.

fine $\mathcal{B}_t$ from the sequence $X(t)$, $t = 0, 1, \cdots$, as above. As a result, $\{X(t), \ t = 0, 1, \cdots, \}$ becomes a stochastic process on the filtered space $(\Omega, \mathcal{F}, \mathcal{B}_t, P)$. If $\{X(t), \ t = 0, 1, \cdots, \}$ is a stochastic process on the filtered space $(\Omega, \mathcal{F}, \mathcal{F}_t, P)$, that is, a time-dependent information structure $\mathcal{F}_t$ is already given, then the natural information structure $\mathcal{B}_t$ contains no more information than $\mathcal{F}_t$, meaning that for each t, $\mathcal{B}_t \subseteq \mathcal{F}_t$. This is because in this case $X(t)$ is a random variable on $(\Omega, \mathcal{F}_t, P)$ and from the dicussion following Example 2.3 in Section 2.2, $\mathcal{B}_t$ is coarser than $\mathcal{F}_t$. Very often, $\mathcal{B}_t$, $t = 0, 1, \cdots$, contains strictly less information than $\mathcal{F}_t$, $t = 0, 1, \cdots$. For example, if $\{X(0), X(1), \cdots, X(t), \cdots, \}$ are the prices of a stock over a certain period of time, $\mathcal{B}_0, \mathcal{B}_1, \cdots, \mathcal{B}_t, \cdots$, represents all the information obtainable from the stock prices, while $\mathcal{F}_0, \mathcal{F}_1, \cdots, \mathcal{F}_t, \cdots$, may represent all the information from the stock market in which this stock is traded and from economic indices as well. To help readers understand this, let us look at the following example.

Example 3.2   Suppose that an economy has 3 possible states, i.e., $\Omega = \{\omega_1, \omega_2, \omega_3\}$. The probability of each state is $1/3$. At time 0, $\mathcal{F}_0 = \{\emptyset, \Omega\}$. At both times 1 and 2, one and only one of the states will prevail. Thus, $\mathcal{F}_1$ and $\mathcal{F}_2$ are the same and is generated by the partition $\{\{\omega_1\}\{\omega_2\}, \{\omega_3\}\}$. Consider now a stock with time-0 price of $103\frac{1}{3}$. Let $S(t)$ be the stock price at time t. Then $S(0) = 103\frac{1}{3}$. Assume that at time 1, $S(1) = 110$, if one of the states ω_1, ω_2 prevails, and $S(1) = 90$, if the state ω_3 prevails. At time 2, $S(2) = 120$ if ω_1 prevails, $S(2) = 100$ if ω_2 prevails, and $S(2) = 90$ if ω_3 prevails. Thus, $\mathcal{B}_1$ is generated by the partition $\{\{\omega_1, \omega_2\}, \{\omega_3\}\}$ and $\mathcal{B}_2 = \mathcal{F}_2$. Obviously, $\mathcal{B}_1 \neq \mathcal{F}_1$. Therefore, the information structure generated by the stock price contains less information than the information structure of the economy.

$\square$

The idea of obtaining an information structure from a sequence of random variables can extend to the case where there are two or more sequences of random variables. For instance, if we have two sequences $\{X(t)\}$ and $\{Y(t)\}$, $t = 0, 1, \cdots$. The information structure at time t, $\mathcal{B}_t$, obtained from $\{X(t)\}$ and $\{Y(t)\}$, will be generated by

$$\mathcal{B}_{X(0)}, \mathcal{B}_{Y(0)}, \mathcal{B}_{X(1)}, \mathcal{B}_{Y(1)}, \cdots, \mathcal{B}_{X(t)}, \mathcal{B}_{Y(t)}.$$

3.2 RANDOM WALKS

Random walks form the simplest class of discrete-time stochastic processes. Because these processes are simple, intuitive and have other appealing features, they have been widely used in modelling financial securities. In this section we show how a random walk is constructed and how it can be used to model financial securities.

We begin with a standard random walk over the period $[0, T]$, which is described as follows.

> Imagine that an object starts at a position marked 0. The time period $[0, T]$ is divided into equal intervals of time steps of size $\tau > 0$. During each time step the object moves either up h units with probability $1/2$ or down h units with probability $1/2$.

Let $X(t)$ denote the position of the object at time t. Then, $X(t), t = 0, \tau, 2\tau, \cdots, T$, is a sequence of time-dependent random variables, which we will call a standard random walk. Note that we have not yet defined the associated information structure. However, from the discussion in the previous section, we may choose the natural information structure generated from $X(t), t = 0, \tau, 2\tau, \cdots, T$. A detailed discussion of $X(t), t = 0, \tau, 2\tau, \cdots, T$, and its generalization are given in the following paragraphs.

Introduce a sequence of independent, identically distributed (iid) binomial random variables $Y_1, Y_2, \cdots, Y_k, \cdots$, on a probability space $(\Omega, \mathcal{F}, P)$ as follows. For a given $h > 0$, let

$$P\{Y_k = h\} = P\{Y_k = -h\} = \frac{1}{2}. \qquad (3.1)$$

Then we have

$$X(t) = Y_1 + Y_2 + \cdots + Y_{\bar{t}}, \qquad (3.2)$$

where $\bar{t} = t/\tau$ is the number of time intervals before time t. Thus, Y_k represents the kth movement of the random walk. For each fixed t, $X(t)$ is binomially distributed. Let m be the number of times the object moves up. Then $m = 0, 1, \cdots, \bar{t}$. The position of the object with m up-moves is $x = mh - (\bar{t} - m)h = (2m - \bar{t})h$ and the corresponding probability is

$$P\{X(t) = x\} = \left(\begin{array}{c} \bar{t} \\ m \end{array} \right) \left[\frac{1}{2} \right]^{\bar{t}}. \qquad (3.3)$$

The mean and variance of $X(t)$ can be calculated easily:

$$E(X(t)) = \bar{t} E(Y_1) = 0, \qquad (3.4)$$

$$Var(X(t)) = \bar{t} Var(Y_1) = \bar{t} h^2 = t\frac{h^2}{\tau}. \qquad (3.5)$$

For $X(t), t = 0, \tau, 2\tau, \cdots, T$, to be a process, a time-dependent information structure is needed. We use the natural information structure $\mathcal{B}_t$ in this case. For each t, $\mathcal{B}_t$ consists of all possible unions of paths up to time t. In other words, the corresponding partition (see Example 2.1) consists of all possible paths up to time t. For example, The partition for $\mathcal{B}_2$ is

$$\{(h, h), (h, -h), (-h, h), (-h, -h)\}.$$

Thus we define a stochastic process $\{X(t)\}$ and call it a *standard random walk*. A probability tree for the standard random walk is given in Figure 3.1 and is called a binomial tree.

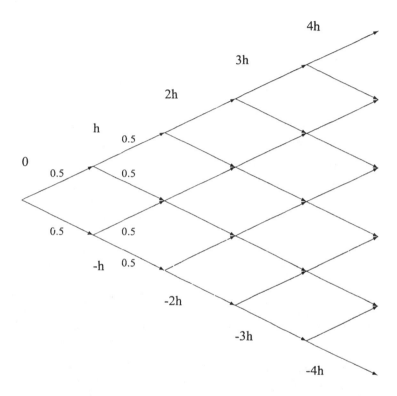

Figure 3.1.   The tree of a standard random walk.

The probability distribution of $X(t)$ can also be calculated recursively in t. A recursive formula is very useful when we use random walks to model financial securities over an extended period of time.

Let $g(x,t) = P\{X(t) = x\}$. Recall that

$$X(t + \tau) = Y_1 + \cdots + Y_{\bar{t}} + Y_{\bar{t}+1} = X(t) + Y_{\bar{t}+1}.$$

It follows from the Law of Total Probability that

$$g(x, t + \tau) = P\{X(t) = x - Y_{\bar{t}+1}\} = E\{P\{X(t) = x - Y_{\bar{t}+1}|Y_{\bar{t}+1}\}\}.$$

Since $X(t)$ and $Y_{\bar{t}+1}$ are independent,

$$P\{X(t) = x - Y_{\bar{t}+1}|Y_{\bar{t}+1} = h\} = g(x - h, t)$$

and

$$P\{X(t) = x - Y_{\bar{t}+1}|Y_{\bar{t}+1} = -h\} = g(x + h, t).$$

One obtains

$$
\begin{aligned}
&E\{P\{X(t) = x - Y_{\bar{t}+1}|Y_{\bar{t}+1}\}\} \\
=\ & g(x - h, t)P\{Y_{\bar{t}+1} = h\} + g(x + h, t)P\{Y_{\bar{t}+1} = -h\} \\
=\ & g(x - h, t)g(h, \tau) + g(x + h, t)g(-h, \tau).
\end{aligned}
$$

Thus

$$g(x, t + \tau) = g(x - h, t)g(h, \tau) + g(x + h, t)g(-h, \tau). \tag{3.6}$$

Obviously, $g(0, 0) = 1$, $g(x, 0) = 0, x \neq 0$.

For the standard random walk, $g(h, \tau) = g(-h, \tau) = \frac{1}{2}$. We obtain

$$g(x, t + \tau) = \frac{1}{2}[g(x - h, t) + g(x + h, t)]. \tag{3.7}$$

As we have seen in (3.4) and (3.5), the mean and variance of the standard random walk are proportional to the time that has passed. In other words, the average mean $\frac{1}{t}E(X(t))$ and the average variance $\frac{1}{t}Var(X(t))$ remain constant over time[3]. Furthermore, the move at time t depends only on the position of the object at time t, not on its past positions. The latter property is called the Markov property. These properties are appealing to financial modelling since some risky securities behave this way and the construction of a tree from a stochastic process with these properties is relatively simple. However, in many financial applications, it is required to construct a random walk with a nonzero drift. In the following we extend the idea discussed above to random walks with constant nonzero drift and constant volatility.

We now wish to construct a random walk $\{X(t),\ t = 0, \tau, 2\tau, \ldots, \}$ such that

$$E(X(t)) = t\mu, \quad Var(X(t)) = t\sigma^2, \tag{3.8}$$

where μ and σ are the given drift and volatility.

There are two ways to achieve this goal.

1. Adjust the probabilities of the up movement and the down movement. Let

$$P\{Y_k = h\} = p, \quad P\{Y_k = -h\} = 1 - p,$$

---

[3] In finance, the average mean and the square root of the average variance of a random walk are often referred to as the drift and the volatility of the random walk.

for all k, where q is to be determined. Recall that

$$E(X(t)) = \frac{t}{\tau}E(Y_1), \text{ and } Var(X(t)) = \frac{t}{\tau}Var(Y_1).$$

Since

$$E(Y_1) = h(2p - 1), \quad Var(Y_1) = 4h^2p(1 - p),$$

to satisfy the equations in (3.8) we must have

$$\frac{h(2p - 1)}{\tau} = \mu, \quad \frac{4h^2p(1 - p)}{\tau} = \sigma^2.$$

Thus

$$h^2(2p - 1)^2 = \mu^2\tau^2, \quad 4h^2p(1 - p) = \sigma^2\tau.$$

Since $(2p - 1)^2 + 4p(1 - p) = 1$, adding the two identities side by side yields $h^2 = \sigma^2\tau + \mu^2\tau^2$, or $h = \sqrt{\sigma^2\tau + \mu^2\tau^2}$. To find q, we replace h in the first identity by $\sqrt{\sigma^2\tau + \mu^2\tau^2}$. Thus

$$p = \frac{1}{2}\left[1 + \frac{\mu\tau}{\sqrt{\sigma^2\tau + \mu^2\tau^2}}\right] = \frac{1}{2}\left[1 + \sqrt{\frac{1}{1 + \sigma^2/\mu^2\tau}}\right].$$

Therefore,

$$h = \sqrt{\sigma^2\tau + \mu^2\tau^2}, \quad p = \frac{1}{2}\left[1 + \sqrt{\frac{1}{1 + \sigma^2/\mu^2\tau}}\right]. \tag{3.9}$$

It follows from (3.6) with $g(h, \tau) = p$ and $g(-h, \tau) = 1 - p$ that the corresponding recursive formula becomes

$$g(x, t + \tau) = pg(x - h, t) + (1 - p)g(x + h, t). \tag{3.10}$$

2. Adjust the sizes of the up movement and the down movement separately.

$$P\{Y_k = h_1\} = \frac{1}{2}, \quad P\{Y_k = -h_2\} = \frac{1}{2}.$$

Since

$$E(Y_k) = \frac{1}{2}(h_1 - h_2), \quad Var(Y_k) = \left(\frac{h_1 + h_2}{2}\right)^2,$$

to satisfy the equations in (3.8) we must have

$$\frac{h_1 - h_2}{2\tau} = \mu, \quad \frac{(h_1 + h_2)^2}{4\tau} = \sigma^2.$$

Thus, we have two linear equations

$$h_1 - h_2 = 2\mu\tau, \quad h_1 + h_2 = 2\sigma\sqrt{\tau}.$$

The second equation is obtained by taking square roots. Solving the equations yields

$$h_1 = \mu\tau + \sigma\sqrt{\tau}, \quad h_2 = -\mu\tau + \sigma\sqrt{\tau}. \tag{3.11}$$

The corresponding recursive formula can be obtained by slightly modifying (3.6). With different sizes for the up-move and the down-move,

$$g(x, t + \tau) = g(x - h_1, t)g(h_1, \tau) + g(x + h_2, t)g(-h_2, \tau). \tag{3.12}$$

Thus, in this case

$$g(x, t + \tau) = \frac{1}{2}[g(x - h_1, t) + g(x + h_2, t)]. \tag{3.13}$$

It is easy to see from these constructions that a random walk (with or without nonzero drift) has the Markov property: future positions of the random walk depend only on the current position of the walk, but not on its past positions. Moreover, a random walk has independent increments, i.e., for any sequence $0 \leq t_1 < t_2 < \cdots < t_n$, the increments

$$X(t_1), \; X(t_2) - X(t_1), \; \cdots, \; X(t_n) - X(t_{n-1})$$

are independent.

Readers may notice that in this section time is measured in 'real' time instead of integer units $0, 1, 2, \cdots$. In many practical situations, trading times are not equally spaced. For instance, when dealing with daily trading, we must take weekends into consideration. Thus, this setting allows for modification of a random walk such that step sizes are adjusted to mimic actual trading times. Also notice that we assume the jump sizes, $Y_1, Y_2, \cdots$, are independent identically distributed two-state binomial random variables. The independence guarantees that the process is Markovian. A further modification is often required to model financial securities more accurately. In many cases, the Markov property is maintained but each Y_k follows a discrete distribution and its value depends on Y_{k-1} only. For example, Y_k may be chosen to be a three-state random variable. As a result, we obtain a trinomial model. Trinomial models are fairly popular in option pricing since they converge faster than their binomial counterparts. In the next section, we further demonstrate how to modify random walks to meet various needs in financial modelling.

Modelling stock price over a longer period of time

We now describe in general how a random walk with a non-zero drift can be used to model the price movement of a stock or a risky security. Consider a stock for the time period $[0, \ T]$. Assume that during period $[0, \ T]$, there are $\bar{T}$ trading dates, $\bar{t} = 0, 1, \cdots, \bar{T} - 1$, separated by regular intervals (trading periods), i.e. $\bar{t} = t/\tau$, where τ is the time between two consecutive trading dates. At each time t, assume[4] that there are only two states of the economy for each trading period: the upstate and the downstate. The probablities of the upstate and the downstate are p and $1 - p$, respectively. The return of the stock for each period is u when the upstate is attained and d when the downstate is attained. Suppose that $S(t)$ is the price of the stock at time t. Define $Y_{\bar{t}} = \ln S(t) - \ln S(t - \tau)$. Then $Y_{\bar{t}}$ is a binomial random variable with

$$P\{Y_{\bar{t}} = \ln u\} = p, \ \ P\{Y_{\bar{t}} = \ln d\} = 1 - p. \tag{3.14}$$

Let $X(t) = Y_1 + Y_2 + \cdots + Y_{\bar{t}}$. Then $\{X(t)\}$ is a random walk and the price process $\{S(t)\}$ can be expressed as

$$S(t) = S(0)e^{X(t)}. \tag{3.15}$$

The process (3.15), i.e., the exponential function of a random walk, is referred to as a geometric random walk or more commonly a binomial tree.

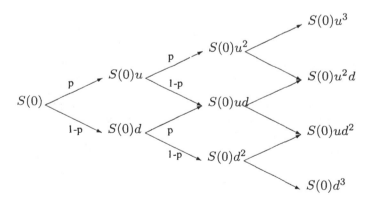

Figure 3.2. The binomial model of the stock price.

[4]The model we consider here is for illustration purposes. Readers may extend the model to meet actual modelling needs. See the earlier comments.

In Figure 3.2., we give the binomial tree of the stock price process in three time steps. The binomial tree shows a very appealing feature of the model: recombining. In a recombining binomial model, the security price is determined only by the number of upstates and the number of downstates that have occurred in the past and is independent of the order in which those states occurred. If a model is recombining, the number of states or nodes increases linearly instead of exponentially. This is critically important in numerical finance since we often need to model a security price over a long period of time.

We further assume that the logarithmic return of $S(t)$ has a constant average mean μ (μ is called the drift of the stock) and a constant average variance σ^2 (σ is called the volatility of the stock), i.e.,

$$E\{\ln S(t)\} = \ln S(0) + \mu t, \quad \text{and} \quad Var\{\ln S(t)\} = \sigma^2 t.$$

Under this assumption, the approach (3.9) in the previous section applies and we obtain

$$u = e^{\sqrt{\sigma^2 \tau + \mu^2 \tau^2}}, \quad d = e^{-\sqrt{\sigma^2 \tau + \mu^2 \tau^2}}. \tag{3.16}$$

and

$$p = \frac{1}{2}\left[1 + \sqrt{\frac{1}{1 + \sigma^2/\mu^2 \tau}}\right]. \tag{3.17}$$

If we use the second approach (3.11) in the previous section, then $p = \frac{1}{2}$ and

$$u = e^{\mu \tau + \sigma \sqrt{\tau}}, \quad d = e^{\mu \tau - \sigma \sqrt{\tau}}. \tag{3.18}$$

Example 3.4   Suppose that we want to model a stock with current price of 100 for a period of one year. Its one-year expected rate of return is estimated as 20% and its volatility during this one-year period is 30%. It is assumed that the stock is traded at the end of each month. Let's further assume that the stock price follows a geometric random walk, i.e., $S(t) = 100e^{X(t)}$, where $\{X(t)\}$ is a random walk. Adapt the second approach in the previous section in which we fix the probabilities to be 1/2 and change the sizes of the up and down movements. As described above, we obtain formula (3.18). In this case, the parameter values are given as follows: $q = 0.5$, $\tau = 1/12$ =one month, and $\sigma = 0.3$. The annual expected return of the stock satisfies

$$E\left\{\frac{S(1)}{S(0)}\right\} = E(e^{X(1)}) = 1.2.$$

Since

$$
\begin{aligned}
E(e^{X(1)}) &= E\left\{e^{Y_1 + \cdots + Y_{12}}\right\} = E\left\{e^{Y_1}\right\} \cdots E\left\{e^{Y_{12}}\right\} \\
&= \left(E\left\{e^{Y_1}\right\}\right)^{12} = \left(\frac{1}{2}[e^{\mu \tau + \sigma \sqrt{\tau}} + e^{\mu \tau - \sigma \sqrt{\tau}}]\right)^{12},
\end{aligned}
$$

we have

$$\left(\frac{1}{2}[e^{\mu\tau+\sigma\sqrt{\tau}} + e^{\mu\tau-\sigma\sqrt{\tau}}]\right)^{12} = 1.2.$$

Solving this equation yields $\mu = 0.13738$. Other parameter values are easily obtained now. We have $u = e^{\mu\tau+\sigma\sqrt{\tau}} = 1.10302$, and $d = e^{\mu\tau-\sigma\sqrt{\tau}} = 0.92760$. In Figure 3.3, we give the stock price process for the first three months.

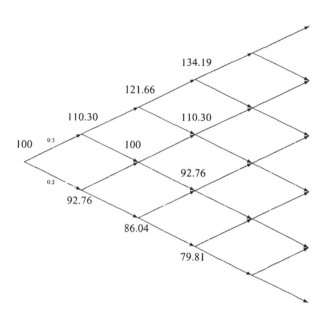

Figure 3.3. The binomial tree of the stock price.

3.3 DISCRETE-TIME MARKOV CHAINS

In this section we introduce a special class of discrete stochastic processes: discrete-time Markov chains. Consider a stochastic process $\{X(0), X(1), \cdots, X(t), \cdots\}$ that takes only a finite or a countable number of values (often called states): $x_0, x_1, \cdots, x_n, \cdots$. Such a process is called a discrete Markov chain if

$$P\{X(t+1) = x_j | X(t) = x_i, X(t-1) = x_h, \cdots, X(0) = x_g\}$$
$$= P\{X(t+1) = x_j | X(t) = x_i\}, \tag{3.19}$$

for any $x_j, x_i, x_h, \cdots, x_g$ and the above probability is independent of t. Intuitively, the distribution of the future value $X(t+1)$ depends on the current value $X(t)$, but not on the past values $X(t-1), \cdots, X(0)$, . the Markov property we described earlier.

Denote

$$p_{ij} = \Pr\{X(t+1) = x_j \mid X(t) = x_i\}, \tag{3.20}$$

the probability that the process moves from the state x_i to the state x_j in one time step. This probability is independent of t due to the Markov property and is called the transition probability from the state x_i to the state x_j. Obviously, for each fixed i, $p_{ij}, j = 0, 1, \cdots$, is a probability distribution and hence $\sum_{j=0}^{\infty} p_{ij} = 1, i = 0, 1, \cdots$. Define a matrix **P** as follows:

$$\mathbf{P} = \begin{bmatrix} p_{00} & p_{01} & \cdots \\ p_{10} & p_{11} & \cdots \\ & \cdots & \end{bmatrix}. \tag{3.21}$$

The matrix **P** is called the transition probability matrix of the Markov chain. The rows of the matrix represent the probability distibutions of the next move when the process is in different states. Clearly, a Markov chain is characterized by its transition probability matrix.

Although we have defined a fairly specific class of stochastic processes, this class contains many stochastic processes used in financial modelling. For example, it contains the random walks discussed in the previous section. To see this, we consider a random walk with jump size h. The possible states of the walk are

$$0, -h, h, -2h, 2h, \cdots.$$

Denote

$$x_0 = 0, x_1 = -h, x_2 = h, x_3 = -2h, x_4 = 2h,$$

and in general,

$$x_{2i-1} = -ih, \quad x_{2i} = ih, \quad i = 1, 2, \cdots.$$

Then, the corresponding probability transition matrix is given by

$$\mathbf{P} = \begin{bmatrix} 0 & 1-p & p & 0 & 0 & 0 & 0 & \cdots \\ p & 0 & 0 & 1-p & 0 & 0 & 0 & \cdots \\ 1-p & 0 & 0 & 0 & p & 0 & 0 & \cdots \\ 0 & p & 0 & 0 & 0 & 1-p & 0 & \cdots \\ 0 & 0 & 1-p & 0 & 0 & 0 & p & \cdots \\ & & \cdots & & & & & \end{bmatrix},$$

where p is the probability that the random walk moves h unites up in one time step.

We now study the probability distribution of a Markov chain after n time steps. Define $p_{ij}(n) = P\{X(t + n) = x_j \mid X(t) = x_i\}$, and

$$\mathbf{P}(n) = \begin{bmatrix} p_{00}(n) & p_{01}(n) & \cdots \\ p_{10}(n) & p_{11}(n) & \cdots \\ & \cdots & \end{bmatrix}. \tag{3.22}$$

$\mathbf{P}(n)$ represents the transition probabilities after n time steps. Under the law of total probability, for any integer m and n,

$$p_{ij}(m + n) = P\{X(t + m + n) = x_j \mid X(t) = x_i\}$$

$$= \sum_{k=0}^{\infty} P\{X(t + m + n) = x_j \mid X(t + m) = x_k, X(t) = x_i\}$$

$$\times \quad P\{X(t + m) = x_k \mid X(t) = x_i\}$$

$$= \sum_{k=0}^{\infty} P\{X(t + m + n) = x_j \mid X(t + m) = x_k\}$$

$$\times \quad P\{X(t + m) = x_k \mid X(t) = x_i\}$$

$$= \sum_{k=0}^{\infty} p_{kj}(n) p_{ik}(m) = \sum_{k=0}^{\infty} p_{ik}(m) p_{kj}(n). \tag{3.23}$$

In matrix notation, equation (3.23) is equivalent to

$$\mathbf{P}(m + n) = \mathbf{P}(m)\mathbf{P}(n). \tag{3.24}$$

Equation (3.23) or (3.24) is called the Chapman-Kolmogorov Equation. The Chapman-Kolmogorov Equation can be used to compute transition probabilities in any number of time steps. Obviously, $\mathbf{P}(1) = \mathbf{P}$. Thus,

$$\mathbf{P}(2) = \mathbf{P}(1)\mathbf{P}(1) = \mathbf{P}^2$$

and

$$\mathbf{P}(3) = \mathbf{P}(2)\mathbf{P}(1) = \mathbf{P}^2\mathbf{P} = \mathbf{P}^3.$$

Continuing in this manner, we have $\mathbf{P}(n) = \mathbf{P}^n$.

We now consider the distribution of $X(t)$ for a fixed t. Suppose that $\Pi(0) = (\pi_0(0), \pi_1(0), \pi_2(0), \cdots)$ is the probability distribution of the initial state $X(0)$. In particular if a Markov chain starts at some state, say x_i, the probability distribution $\Pi(0)$ is a singleton, i.e. $\pi_i(0) = 1$ and $\pi_j(0) = 0$ for any $j \neq i$. For the initial probability distribution $\Pi(0) = (\pi_0(0), \pi_1(0), \pi_2(0), \cdots)$, the probability distribution of $X(t)$ can be computed as follows.

$$P\{X(t) = x_j\} = \sum_{i=0}^{\infty} P\{X(t) = x_j \mid X(0) = x_i\} P\{X(0) = x_i\}$$

$$= \sum_{i=0}^{\infty} p_{ij}(t)\pi_i(0). \tag{3.25}$$

In other words, the probability distribution of $X(t)$ as a row vector is

$$\Pi(0)\mathbf{P}(t) = \Pi(0)\mathbf{P}^t. \tag{3.26}$$

For example, the distribution of $X(1)$ is $\Pi(0)\mathbf{P}$ and the distribution of $X(2)$ is $\Pi(0)\mathbf{P}^2$. The relation between the distributions of $X(s)$ and $X(t)$ for $s > t$ can be derived similarly and is given as

$$P\{X(s) = x_j\} = \sum_{i=0}^{\infty} p_{ij}(s - t)P\{X(t) = x_i\}. \tag{3.27}$$

For many Markov chains, the distribution of $X(t)$ will approach a limiting distribution independent of the initial state, as t goes to infinity. Mathematically, this implies that $\lim_{t \to \infty} p_{ij}(t)$ exists and is independent of i. Denote

$$\pi_j = \lim_{t \to \infty} p_{ij}(t), \text{ and } \Pi = (\pi_0, \pi_1, \cdots, \pi_j, \cdots). \tag{3.28}$$

Then, Π is a probability distribution and π_j roughly represents the proportion of time that a Markov chain stays at x_j over an extended period of time. The distribution Π can be computed easily using the Chapman-Kolmogorov Equation. From (3.23),

$$p_{ij}(t + 1) = \sum_{k=0}^{\infty} p_{ik}(t)p_{kj}.$$

Letting $t \to \infty$, we have

$$\pi_j = \sum_{k=0}^{\infty} \pi_k p_{kj}, \tag{3.29}$$

which is equivalent to

$$\Pi = \Pi\mathbf{P}. \tag{3.30}$$

Hence, the limiting distribution of $\{X(t)\}$ is the solution of the system of linear equations (3.30). The distribution Π can also be viewed as a steady-state distribution of the Markov chain. To see this, suppose that the probability distribution of $X(t)$ is Π for some t. Then, by (3.27) Π is the probability distribution of $X(t + 1)$. Since $X(t + 1)$ has probability distribution Π, so does $X(t + 2)$. Continuing in this way, we have that Π is the probability distribution of $X(s)$ for all $s \geq t$.

Example 3.6 Bond Valuation with Default Risk
Suppose that there are four possible credit ratings in the bond market:

A   for bonds issued by a high quality firm;
B   for bonds issued by a medium quality firm;
C   for bonds issued by a low quality firm;
D   for bonds which are in default.

Suppose that the transition probability matrix of credit ratings over a one-year period[5] is given by

$$\mathbf{P} = \begin{bmatrix} & A & B & C & D \\ A & 0.9601 & 0.0153 & 0.0246 & 0 \\ B & 0.0195 & 0.8502 & 0.1085 & 0.0218 \\ C & 0.0103 & 0.0946 & 0.6905 & 0.2046 \\ D & 0 & 0 & 0 & 1 \end{bmatrix}.$$

For example, the probability that an A-rated bond retains its rating after one year is 0.9601 and the probability that it becomes a B-rated bond is 0.0153. The transition probability matrices for two years and three years are

$$\mathbf{P}^2 = \begin{bmatrix} 0.9223 & 0.0300 & 0.0423 & 0.0054 \\ 0.0364 & 0.7334 & 0.1677 & 0.0625 \\ 0.0188 & 0.1459 & 0.4874 & 0.3479 \\ 0 & 0 & 0 & 1 \end{bmatrix},$$

and

$$\mathbf{P}^3 = \begin{bmatrix} 0.8866 & 0.0436 & 0.0551 & 0.0147 \\ 0.0510 & 0.6400 & 0.1962 & 0.1128 \\ 0.0260 & 0.1704 & 0.3528 & 0.4508 \\ 0 & 0 & 0 & 1 \end{bmatrix}.$$

Assume that there are two zero-coupon bonds: an A-rated bond A and a B-rated bond B. Both bonds have principal of 100 and mature at the end of three years. Bond A will pay an annual interest rate of 4% while bond B will pay an annual interest rate of 4.5%, 50 basis points higher because of a lower rating. The recovery rate for both bonds is assumed to be 60%, which means that the issuer will pay 60% of the principal in case of default. To evaluate each bond, we calculate its expected payoff. A general formula for the expected payoff of a bond at maturity is given by

$$\text{Expected Payoff} = \Pi(0)\mathbf{P}^T\beta, \qquad (3.31)$$

where $\Pi(0)$ is the probability distribution corresponding to the rating of a bond, T is the time to maturity, and β is the payoff pattern of the bond at maturity. In this example, we have $T = 3$, $\Pi_A(0) = (1, 0, 0, 0)$ for bond A

---

[5]We only consider four rating classes for illustration purposes. The actual rating classes are AAA, AA, A, BBB, BB, B, CCC and D and the transition probability matrix of credit ratings can be obtained from a rating agency such as Standard and Poor or Moody's Investors Service.

and $\Pi_B(0) = (0, 1, 0, 0)$ for bond B. The payoff patterns for bonds A and B are

$$\beta_A = 100 \begin{bmatrix} (1+0.04)^3 \\ (1+0.04)^3 \\ (1+0.04)^3 \\ 0.6 \end{bmatrix}, \quad \beta_B = 100 \begin{bmatrix} (1+0.045)^3 \\ (1+0.045)^3 \\ (1+0.045)^3 \\ 0.6 \end{bmatrix},$$

respectively. Thus, the expected payoff for bond A is 111.72 and the expected payoff for bond B is 108.11. The two bonds would be comparable if their expected payoffs were the same. This would require increasing the interest rate on bond B while holding that of bond A 4%.

□

3.4 MARTINGALES AND CHANGE OF PROBABILITY MEASURE

The concept of martingale plays a vital role in the pricing of derivative securities. A stochastic process $\{M(t), t = 0, 1, 2, \cdots, \}$ with $E\{(|M(t)|\} < \infty$ for all t on $(\Omega, \mathcal{F}, \mathcal{F}_t, P)$ is called a martingale if

$$E\{M(t+1) \,|\mathcal{F}_t\} = M(t). \tag{3.32}$$

An intuitive interpretation of Equation (3.32) is that given the current information, one can not predict the value of $M(t+1)$. If $M(t)$ represents the winnings at time t in a series of games, then the martingale property says that this game series is fair to both players.

Equivalent forms of (3.32) are often used to verify martingales. Recalling that $M(t)$ is a random variable on $\mathcal{F}_t$, (2.88) implies that $E\{M(t)|\mathcal{F}_t\} = M(t)$. Thus,

$$\begin{aligned} E\{M(t+1) - M(t) \,|\mathcal{F}_t\} &= E\{M(t+1)|\mathcal{F}_t\} - E\{M(t)|\mathcal{F}_t\} \\ &= E\{M(t+1)|\mathcal{F}_t\} - M(t) = 0. \end{aligned}$$

We obtain an equivalent form of (3.32)

$$E\{M(t+1) - M(t) \,|\mathcal{F}_t\} = 0. \tag{3.33}$$

If $M(t) > 0$, it follows from (3.32) that

$$E\{M(t+1)/M(t) \,|\mathcal{F}_t\} = \frac{1}{M(t)} E\{M(t+1)|\mathcal{F}_t\} = \frac{1}{M(t)} M(t) = 1.$$

We obtain another equivalent form of (3.32)

$$E\{M(t+1)/M(t) \,|\mathcal{F}_t\} = 1, \tag{3.34}$$

if $M(t) > 0$. In many situations, (3.33) or (3.34) is easier than (3.32) to verify.

Some basic properties follow immediately from the martingale definition.
(i) For any t, $E\{M(t)\} = M(0)$, i.e., the expectation of a martingale remains
constant over time.
(ii) For any random variable Y on $\mathcal{F}_T$ with $E\{|Y|\} < \infty$, the stochastic
process $\{ M(t) = E(Y|\,\mathcal{F}_t) \}$ is a martingale.
Property (i) follows by taking expectations in (3.32) and from the Law of
Iterated Expectation. Property (ii) is the immediate result of the Law of
Iterated Expectation.

In many cases, the information structure $\mathcal{F}_t$ is replaced by the natural
information structure $\mathcal{B}_t$ introduced in Section 3.1. In this case (3.32) may be
written as

$$E\{M(t+1)\,|M(t), M(t-1), \cdots, M(0)\} = M(t). \qquad (3.35)$$

If $\{M(t)\}$ is a Markov chain, equation (3.35) reduces to

$$E\{M(t+1)\,|M(t)\} = M(t). \qquad (3.36)$$

It is important to understand that the martingale property is information structure specific. A stochastic process may be a martingale with respect to one
information structure but not with respect to another. The martingale property, however, is preserved with respect to a coarser information structure.
To see this, we consider the stochastic process $\{M(t)\}$ on $(\Omega, \mathcal{F}, \mathcal{F}_t, P)$. As
discussed in Section 3.1, it is also a stochastic process on $(\Omega, \mathcal{F}, \mathcal{B}_t, P)$. First
assume that $\{M(t)\}$ is a martingale with respect to $\mathcal{F}_t$, i.e.,

$$E\{M(t+1)|\mathcal{F}_t\} = M(t).$$

Since $\mathcal{B}_t \subset \mathcal{F}_t$, we have, by the Law of Iterated Expectation (2.92),

$$E\{M(t+1)|\mathcal{B}_t\} = E\{E\{M(t+1)|\mathcal{F}_t\}|\mathcal{B}_t\} = E\{M(t)|\mathcal{B}_t\} = M(t).$$

Thus, $\{M(t)\}$ is a martingale with respect to $\mathcal{B}_t$. Conversely, suppose that
$\{M(t)\}$ is a martingale with respect to $\mathcal{B}_t$. Then

$$E\{M(t+1)|\mathcal{B}_t\} = M(t).$$

If $\mathcal{B}_t$ is strictly coarser than $\mathcal{F}_t$, $E\{M(t+1)|\mathcal{F}_t\}$ might not be equal to
$E\{M(t+1)|\mathcal{B}_t\}$, in which case

$$E\{M(t+1)|\mathcal{F}_t\} \neq M(t),$$

so that $\{M(t)\}$ is not a martingale with respect to $\mathcal{F}_t$. This becomes more
clear in the following example.

Example 3.7 (Continuation of Example 3.2) Recall that in example 3.2, we
consider a stock in a 3-state economy, $\{\omega_1, \omega_2, \omega_3\}$. The probability that

each state occurs is $1/3$. At time 0, $\mathcal{F}_0 = \{\emptyset, \Omega\}$. At both times 1 and 2, one and only one of the states will prevail, i.e., $\mathcal{F}_1$ and $\mathcal{F}_2$ are the same and their corresponding partition is $\{\{\omega_1\}, \{\omega_2\}, \{\omega_3\}\}$. The time-0 price of the stock is $S(0) = 103\frac{1}{3}$. At time 1, $S(1) = 110$, if ω_1 or ω_2 prevails, and $S(1) = 90$, if ω_3 prevails. At time 2, $S(2) = 120$ if ω_1 prevails, $S(2) = 100$ if ω_2 prevails, and $S(2) = 90$ if ω_3 prevails. Further, $\mathcal{B}_1$ is generated by the partition $\{\{\omega_1, \omega_2\}, \{\omega_3\}\}$ and $\mathcal{B}_2 = \mathcal{F}_2$. It follows from (2.91) that on $\{\omega_1, \omega_2\}$,

$$E\{S(2)|\mathcal{B}_1\} = E\{S(2)|\{\omega_1, \omega_2\}\} = 120 \cdot \frac{1}{2} + 100 \cdot \frac{1}{2} = 110 = S(1).$$

Similarly at $\{\omega_3\}$,

$$E\{S(2)|\mathcal{B}_1\} = E\{S(2)|\{\omega_3\}\} = 90 = S(1).$$

It is easy to see that $E\{S(1)\} = 110 \cdot \frac{2}{3} + 90 \cdot \frac{1}{3} = 103\frac{1}{3} = S(0)$. Hence, $S(t)$ is a martingale with respect to $\mathcal{B}_t$. However, $S(t)$ is not a martingale with respect to $\mathcal{F}_t$, since

$$E\{S(2)|\mathcal{F}_1\} = E\{S(2)|\mathcal{F}_2\} = S(2) \neq S(1).$$

$\square$

Let us now revisit the standard random walk $\{X(t)\}$ discussed in section 3.2. For simplicity, and without loss of generality, we assume $\tau = 1$. Thus, $\{X(t) = Y_1 + Y_2 + \cdots + Y_t,\}$ where $P\{Y_t = h\} = P\{Y_t = -h\} = 1/2$. Hence

$$E\{X(t+1) - X(t)\,|X(t)\} = E(Y_{t+1}) = \frac{1}{2}(h - h) = 0.$$

This implies that the standard random walk is a martingale. In the case of the standard geometric random walk

$$S(t) = S(0)e^{X(t)}, \tag{3.37}$$

we have

$$E\{S(t+1)/S(t)|S(t)\} = E(e^{Y_{t+1}}) = \frac{1}{2}(e^h + e^{-h}) > 1.$$

Thus, the standard geometric random walk is not a martingale.

Many financial applications require that a discounted price process be a martingale under a certain probability measure as we will see in Sections 2.6 and 2.7. Hence, if the process is not a martingale under a given probability measure, that measure needs to be replaced with an equivalent one under which the process is a martingale. Such a probability measure is often referred to as a martingale measure. In the following, we show how to find a martingale measure for a stock whose price follows a geometric random walk and extend the method to a more general case in the context of finance.

Let $\{S(t)\}$ be the price of a stock and assume that $\{S(t)\}$ follows a geometric random walk with drift μ and volatility σ. Thus, $S(t) = S(0)e^{X(t)}$, where $X(t) = Y_1 + Y_2 + \cdots + Y_t$, and $Y_t, t = 1, 2, \cdots$, are iid binomial random variables with

$$P\{Y_t = h_1\} = P\{Y_t = -h_2\} = \frac{1}{2}.$$

The values of h_1 and h_2 are given in (3.11). Further let R be the interest rate for each time interval $[t, t + 1]$. Thus, $S^*(t) = (1 + R)^{-t}S(t)$ represents the present or discounted value of the stock price process. If $R = 0$, $S^*(t) = S(t)$, a geometric random walk. We now illustrate how to change the probability measure to a martingale measure Q under which the present value process $S^*(t)$ is a martingale. Suppose that under Q

$$P_Q\{Y_t = h_1\} = q, \quad P_Q\{Y_t = -h_2\} = 1 - q, \tag{3.38}$$

where $P_Q(\cdot)$ denotes probability under the martingale measure Q. Then, from (3.34), $\{S^*(t)\}$ is a martingale if

$$E_Q\{S^*(t + 1)/S^*(t) \mid S^*(t)\} = 1,$$

where E_Q denotes the expectation under the martingale measure Q. Since $S^*(t + 1)/S^*(t) = (1 + R)^{-1}e^{Y_{t+1}}$, the equation above implies

$$uq + d(1 - q) = 1 + R, \tag{3.39}$$

where $u = e^{h_1}$ and $d = e^{-h_2}$. Solving (3.39), we obtain

$$q = \frac{1 + R - d}{u - d}. \tag{3.40}$$

Consider now a more general case where Y_t, $t = 1, 2, \cdots$, are independent but arbitrary random variables.

Assume that the moment generating function $M_{Y_t}(z)$ of Y_t exists for all t. Define, for any real value α_t,

$$Q(t) = e^{\alpha_t Y_t}/M_{Y_t}(\alpha_t).$$

Then, $Q(t), t = 1, 2, \cdots$, are independent, and $E\{Q(t)\} = 1$. Further define

$$Z(t) = Q(1)Q(2) \cdots Q(t). \tag{3.41}$$

$\{Z(t)\}$ is a strictly positive martingale under the old measure since

$$E\{Z(t+1)/Z(t)|Z(t)\} = E\{Q(t+1)\} = 1.$$

Now for any event A at time t, define the probability of A as

$$P_Q(A) = E\{\mathbf{I}_A Z(t)\}. \qquad (3.42)$$

It can be shown that the probability measure $P_Q(\cdot)$, referred to as measure Q, is well and uniquely defined[6]. Since the real numbers α_t are arbitrary, we obtain a family of probability measures. These probability measures are called Esscher measures in the actuarial literature[7].

The relations between the expectation with respect to each measure are given by

$$E_Q\{f(X(t))\} = E\{f(X(t))Z(t)\}, \qquad (3.43)$$

and

$$E_Q\{f(X(s))|X(t)\} = E\{f(X(s))\frac{Z(s)}{Z(t)}|X(t)\}, \ s > t, \quad (3.44)$$

where $f(x)$ is a deterministic function. Thus, for unconditional probabilities or expectations, $Z(t)$ serves as an adjustment factor. If we deal with conditional probabilities or expectations at time t, the corresponding adjustment factor is $Z(s)/Z(t)$.

We now choose the value of α_t for each t such that

$$E\{(1+R)^{-1}e^{Y_t}Q(t)\} = (1+R)^{-1}\frac{M_{Y_t}(\alpha_t + 1)}{M_{Y_t}(\alpha_t)} = 1. \qquad (3.45)$$

Then,

$$\begin{aligned} E_Q\left\{\frac{S^*(t+1)}{S^*(t)}|S^*(t)\right\} &= E\left\{\frac{S^*(t+1)}{S^*(t)}\frac{Z(t+1)}{Z(t)}|S^*(t)\right\} \\ &= (1+R)^{-1}E\{e^{Y_t}Q(t)\} = 1. \end{aligned}$$

Therefore the present value process $\{S^*(t)\}$ is a martingale under probability measure Q.

Example 3.8 Stock Price as a Lognormal Process
In this example, we illustrate how to use the Esscher measures to find a

---

[6]In fact, any strictly positive martingale $\{Z(t)\}$ with $Z(0) = 1$ can be used to create a new probability measure through (3.42). The relations (3.44) will still hold. This idea will be discussed again in Chapter 6.
[7]F. Esscher was a Swedish actuary who proposed a transformation of a distribution using exponential functions. The use of the Esscher measures to find a martingale measure was first proposed by Gerber and Shiu (1996).

martingale measure when a stock price $\{S(t)\}$ is assumed to be a lognormal process, i.e. for each t, $S(t)$ is lognormally distributed (Example 2.7).

As described in (3.37), $S(t) = S(0)e^{X(t)}$, where $X(t) = Y_1+Y_2+\cdots+Y_t$. We assume that the Y_t's are independent and normally distributed with mean μ and standard deviation σ. Thus the stock price $S(t)$ at time t follows a lognormal distribution. For notational convenience, we now assume r to the risk-free interest rate compounded continuously or the force of interest, i.e. $e^r = 1 + R$.

To find a martingale measure, we compute the values α_t. The moment generating function of Y_t is given by $M(z) = e^{\mu z+\frac{1}{2}\sigma^2 z^2}$. Since the Y_t's are iid, $\alpha_t = \alpha$, independent of t. Thus

$$Q(t) = \frac{e^{\alpha Y_t}}{e^{\mu\alpha+\frac{1}{2}\sigma^2\alpha^2}}.$$

Equation (3.45) yields

$$
\begin{aligned}
e^r &= \frac{e^{\mu(\alpha+1)+\frac{1}{2}\sigma^2(\alpha+1)^2}}{e^{\mu\alpha+\frac{1}{2}\sigma^2\alpha^2}} \\
&= e^{\mu+\frac{1}{2}\sigma^2(2\alpha+1)}.
\end{aligned}
$$

Thus, $\alpha = \frac{r-\mu-\frac{1}{2}\sigma^2}{\sigma^2}$, and

$$Z(t) = Q(1)Q(2)\cdots Q(t) = \frac{e^{\alpha(Y_1+Y_2+\cdots+Y_t)}}{e^{(\mu\alpha+\frac{1}{2}\sigma^2\alpha^2)t}} = \frac{e^{\alpha X(t)}}{M_{X(t)}(\alpha)}.$$

We now identify the distribution of the stock price process under the martingale measure. To do so we compute the moment generating function of Y_t under the martingale measure.

$$
\begin{aligned}
E_Q\{e^{zY_t}\} &= E\{e^{zY_t}Q(t)\} = \frac{E\{e^{(z+\alpha)Y_t}\}}{e^{\mu\alpha+\frac{1}{2}\sigma^2\alpha^2}} \\
&= \frac{e^{\mu(z+\alpha)+\frac{1}{2}\sigma^2(z+\alpha)^2}}{e^{\mu\alpha+\frac{1}{2}\sigma^2\alpha^2}} = e^{(\mu+\alpha\sigma^2)z+\frac{1}{2}\sigma^2 z^2} \\
&= e^{(r-\frac{1}{2}\sigma^2)z+\frac{1}{2}\sigma^2 z^2}.
\end{aligned}
$$

Hence, Y_t is normally distributed with mean $r - \frac{1}{2}\sigma^2$ and standard deviation σ. Therefore the stock price process is again a lognormal process with a new drift r and the same volatility σ. An interesting fact is that the new drift is independent of the old drift μ.

□

The method of changing probability measures using a martingale is typical in stochastic calculus. Note that for a probability space $(\Omega, \mathcal{F}, P)$ and a given positive random variable Z with $E(Z) = 1$ (Z can be any positive random variable), $P_Q(A) = E\{\mathbf{I}_A Z\}$, termed as measure Q, gives a new probability measure on the space $(\Omega, \mathcal{F})$. The random variable Z is referred to as the Radon-Nikodym derivative of measure Q with respect to measure P and is denoted as $\frac{dQ}{dP}$. Sometimes, one must find a new probability measure for a filtered space $(\Omega, \mathcal{F}, \mathcal{F}_t, P)$ such that a stochastic process associated with this space satisfies some desirable properties (the martingale property, for example). In this case, a positive random variable with unit mean on $(\Omega, \mathcal{F})$ is often not available and we in turn look for a positive martingale process $\{Z(t)\}$ with $E\{Z(t)\} = 1$ for all t. The formula (3.42) thus assigns a new probability for each event in $\mathcal{F}_t$ and hence we define a new probability measure for all events in $\cup_{t \geq 0} \mathcal{F}_t$. However, one needs to check whether or not the probability measure generated in this way is well defined. This is because for any pair $t_1 < t_2$ and any event $A \in \mathcal{F}_{t_1}$, since $A \in \mathcal{F}_{t_2}$ there are two ways to assign a probability to A: one by $P_Q(A) = E\{\mathbf{I}_A Z(t_1)\}$ and the other by $P_Q(A) = E\{\mathbf{I}_A Z(t_2)\}$. A well defined probability measure requires $E\{\mathbf{I}_A Z(t_1)\} = E\{\mathbf{I}_A Z(t_2)\}$. The martingale requirement on $\{Z(t)\}$ actually ensures that the above equality holds. To see this, one has

$$
\begin{aligned}
E\{\mathbf{I}_A Z(t_2)\} &= E\{E\{\mathbf{I}_A Z(t_2)|\mathcal{F}_{t_1}\}\} \\
&= E\{\mathbf{I}_A E\{Z(t_2)|\mathcal{F}_{t_1}\}\} = E\{\mathbf{I}_A Z(t_1)\}.
\end{aligned}
$$

The first equality follows from the law of iterated expectation with respect to information structure, the second equality holds since $\mathbf{I}_A$ is a random variable on $\mathcal{F}_{t_1}$ (see (2.90)), and the third equality follows from the martingale property of $\{Z(t)\}$.

3.5 STOPPING TIMES

Let $(\Omega, \mathcal{F}, \mathcal{F}_t, P)$ be a filtered space. We still assume that $t = 0, 1, 2, \cdots$. A stopping time $\mathcal{T}$ is a counting random variable (i.e. the values of $\mathcal{T}$ are nonnegative integers) on the probability space $(\Omega, \mathcal{F}, P)$ such that for any t, the event $\{\mathcal{T} \leq t\}$ belongs to $\mathcal{F}_t$. We assume that the values of $\mathcal{T}$ are in the set of extended nonnegative numbers, including ∞. If $P\{\mathcal{T} = \infty\} = 0$, we say that $\mathcal{T}$ is finite.

In general, $\mathcal{T}$ is not a random variable on $(\Omega, \mathcal{F}_t, P)$ for a fixed t. Recall that $\mathcal{F}_t$ represents the information up to time t. $\{\mathcal{T} \leq t\} \in \mathcal{F}_t$ means that the "decision" to stop (or not to stop) by time t is based only on the information available up to time t.

To further understand the concept of stopping time, we look at the following example.

Example 3.9 (Continuation of Example 3.7)
Recall that in Example 3.6 an economy has 3 possible states, i.e. $\Omega =$

$\{\omega_1, \omega_2, \omega_3\}$. A stock with price $S(t)$ at time t has the following prices: At time 0, $S(0) = 103\frac{1}{3}$. At time 1, $S(1) = 110$, if ω_1 or ω_2 prevails, and $S(1) = 90$, if ω_3 prevails. At time 2, $S(2) = 120$ if ω_1 prevails, $S(2) = 100$ if ω_2 prevails, and $S(2) = 90$ if ω_3 prevails. We adopt the natural information structure $\mathcal{B}_t$ obtained from the stock price process. As in Example 3.6, $\mathcal{B}_1$ is generated by the partition $\{\{\omega_1, \omega_2\}, \{\omega_3\}\}$, and $\mathcal{B}_2$ is generated by the partition $\{\{\omega_1\}, \{\omega_2\}, \{\omega_3\}\}$.

We now set two stopping rules.

Rule # 1: Buy the stock when its price is 100 or lower;

Rule # 2: Buy the stock at time 1 when the state $\{\omega_2\}$ prevails, and buy the stock at time 2 otherwise.

Let T_1 and T_2 be the time of purchase under Rule # 1 and Rule # 2, respectively. Then T_1 is a stopping time but T_2 is not.

To see this, we examine the event $\{T_i \leq t\}, i = 1, 2$, for $t = 0, 1, 2$. It is easy to check that $\{T_1 = 1\} = \{\omega_3\}$ and $\{T_1 = 2\} = \{\omega_2\}$. Hence, $\{T_1 \leq 0\} = \emptyset, \{T_1 \leq 1\} = \{\omega_3\}$, and $\{T_1 \leq 2\} = \{\omega_2, \omega_3\}$. By definition, T_1 is a stopping time. For T_2, we have $\{T_2 = 1\} = \{\omega_2\}$ and $\{T_2 = 2\} = \{\omega_1, \omega_3\}$. Obviously, $\{T_2 \leq 1\} = \{\omega_2\}$ is not in $\mathcal{B}_1$. Hence T_2 is not a stopping time. It is clear that under Rule # 2, the decision on whether to buy the stock depends on what will occur in the future (time 2), which violates the condition of stopping time.

$\square$

In this discrete-time setting, the following equivalent definition of a stopping time is more convenient: T is a stopping time if and only if for each $t = 0, 1, 2, \cdots, \{T = t\}$ belongs to $\mathcal{F}_t$. The equivalence follows from

$$\{T \leq t\} = \cup_{s=0}^{t}\{T = s\} \quad \text{and} \quad \{T = t\} = \{T \leq t\} - \{T \leq t - 1\}.$$

We now discuss some properties of stopping times. A nonnegative constant $T = c$ is a stopping time, since the event $\{T \leq t\}$ is Ω when $c \leq t$ and $\emptyset$ when $c > t$. If T_1 and T_2 are two stopping times, then the maximum, $T_1 \vee T_2 \overset{def}{=} \max(T_1, T_2)$, and the minimum, $T_1 \wedge T_2 \overset{def}{=} \min(T_1, T_2)$, are also stopping times. In particular, a stopping time truncated from below, $T \vee t$, and a stopping time truncated from above, $T \wedge t$, are stopping times. An important class of stopping times are barrier hitting times or first passage times of stochastic processes. Let $\{X(t)\}$ be a stochastic process. For a real number $b > X(0)$, define

$$T_b = \inf\{t; \ X(t) \geq b\}, \tag{3.46}$$

and for $b < X(0)$, define

$$T_b = \inf\{t; \ X(t) \leq b\}.$$

The random time T_b is the time the process $\{X(t)\}$ hits or passes the barrier $x = b$ for the first time. Since at any time t, we can observe whether or not the process $\{X(t)\}$ hits or passes this barrier, T_b is a stopping time. This can also be seen from

$$\{T_b \leq t\} = \cup_{s=0}^{t}\{X(s) \geq b\},$$

which says that the stopping time T_b is bounded by t if and only if the process $\{X(s)\}$ hits or passes the barrier $x = b$ by time t.

One of the most important results on martingales is the Optional Sampling Theorem. As we will see later, the Optional Sampling Theorem provides a computational tool for the distribution of many barrier hitting times.

Let $\{X(t), t = 0, 1, 2, \cdots, \}$ be a stochastic process and T a stopping time. Then, $X(T)$ is a random variable. Moreover, $\{X(T \wedge t), t = 0, 1, 2, \cdots, \}$ defined by

$$X(T \wedge t) = \left\{ \begin{array}{ll} X(T), & T < t \\ X(t), & T \geq t, \end{array} \right.$$

is a stochastic process and is called the stopped process by T.

The Optional Sampling Theorem (Discrete Version)

If $\{X(t), t = 0, 1, 2, \cdots, \}$ is a martingale, then $\{X(T \wedge t), t = 0, 1, 2, \cdots, \}$ is also a martingale. Furthermore, if $|X(T \wedge t)|$ is bounded for all $t = 0, 1, 2, \cdots$, i.e. there is a constant K independent of t such that $|X(T \wedge t)| \leq K$, then

$$E\{X(T)\} = E\{X(0)\}. \tag{3.47}$$

The proof of ths result involves some advanced mathematics in measure theory and is omitted here. In the following example we illustrate the use of the Optional Sampling Theorem to compute the distribution of a hitting time.

Example 3.10 Barrier Hitting Time of a Geometric Random Walk
As described in Section 3.3, let $S(t) = S(0)e^{X(t)}$ be a geometric random walk, where $X(t) = Y_1 + Y_2 + \cdots + Y_t$, and $P\{Y_t = h\} = p$, $P\{Y_t = -h\} = 1 - p$. Let $U = S(0)e^{mh}$, where m is a positive integer, be an upper barrier of $S(t)$. Define a stopping time T_U as

$$T_U = \inf\{t, \ S(t) = U\}, \tag{3.48}$$

the first time the geometric random walk $S(t)$ hits the barrier U. We remark that from the choice of U, $S(t)$ always hits the barrier when crossing the barrier. Thus $\inf\{t, \ S(t) \geq U\} = \inf\{t, \ S(t) = U\}$.

We now use the Optional Sampling Theorem to compute the Laplace transform of the first passage time T_U.

For any fixed real value $z \geq 0$, consider the process $Z(t) = e^{-zt}[S(t)]^{\xi}$, where ξ is a constant to be determined later. We first seek a value of ξ such that $Z(t)$ is a martingale. Equation (3.34) yields

$$E\{Z(t+1)/Z(t)|Z(t)\} = E\{e^{-z}e^{\xi Y_{t+1}}\} = 1.$$

Equivalently,

$$pe^{\xi h} + (1-p)e^{-\xi h} = e^z.$$

Let $y = e^{\xi h}$. We have

$$py^2 - e^z y + (1-p) = 0.$$

Thus

$$e^{\xi h} = \frac{e^z \pm \sqrt{e^{2z} - 4p(1-p)}}{2p},$$

or

$$\xi = \frac{1}{h}\ln\left(\frac{e^z \pm \sqrt{e^{2z} - 4p(1-p)}}{2p}\right).$$

Since $S(t)$ is always nonnegative but may approach zero, we need to choose ξ to be nonnegative in order to satisfy the condition in the Optional Sampling Theorem. Thus

$$e^{\xi h} = \frac{e^z + \sqrt{e^{2z} - 4p(1-p)}}{2p}. \tag{3.49}$$

With this choice of ξ, $Z(t) = e^{-zt}[S(t)]^\xi$ is a martingale. Since $|Z(T_U \wedge t)| \le U^\xi$, the condition in the Optional Sampling Theorem is satisfied. Therefore we have

$$E\{Z(T_U)\} = Z(0),$$

or

$$E\{e^{-zT_U}[S(T_U)]^\xi\} = [S(0)]^\xi. \tag{3.50}$$

The stopping time T_U may be infinite with positive probability, i.e. $P\{T_U = \infty\} > 0$. We need to take this into consideration when calculating $E\{e^{-zT_U}[S(T_U)]^\xi\}$. As in Section 2.2, let $\mathbf{I}_{\{T_U < \infty\}}$ and $\mathbf{I}_{\{T_U = \infty\}}$ be the indicators for the events $\{T_U < \infty\}$ and $\{T_U = \infty\}$, respectively. Then

$$\mathbf{I}_{\{T_U < \infty\}} + \mathbf{I}_{\{T_U = \infty\}} = 1.$$

Thus

$$\begin{aligned}
&E\{e^{-zT_U}[S(T_U)]^\xi\}\\
&= E\{e^{-zT_U}[S(T_U)]^\xi(\mathbf{I}_{\{T_U < \infty\}} + \mathbf{I}_{\{T_U = \infty\}})\}\\
&= E\{e^{-zT_U}[S(T_U)]^\xi\mathbf{I}_{\{T_U < \infty\}}\} + E\{e^{-zT_U}[S(T_U)]^\xi\mathbf{I}_{\{T_U = \infty\}}\}.
\end{aligned}$$

The second term above is zero since

$$E\{e^{-z \cdot T_U} [S(T_U)]^\xi \, \mathbf{I}_{\{T_U = \infty\}}\} \le E\{e^{-\infty} U^\xi \mathbf{I}_{\{T_U = \infty\}}\} = 0.$$

Thus

$$E\{e^{-zT_U} [S(T_U)]^\xi\} = E\{e^{-zT_U} [S(T_U)]^\xi \mathbf{I}_{\{T_U < \infty\}}\}.$$

Noting that $S(T_U) = U$ for $T_U < \infty$, it follows from (3.50)

$$E\{e^{-zT_U} \mathbf{I}_{\{T_U < \infty\}}\} = \left(\frac{S(0)}{U}\right)^\xi = e^{-m\xi h}. \tag{3.51}$$

With (3.49), we have

$$E\{e^{-zT_U} \mathbf{I}_{\{T_U < \infty\}}\} = \left(\frac{2p}{e^z + \sqrt{e^{2z} - 4p(1-p)}}\right)^m. \tag{3.52}$$

Thus we obtain the (defective) Laplace transform[8] of T_U.

Many properties of the barrier hitting time T_U can be obtained from (3.52). Let $z = 0$. Then,

$$P\{T_U < \infty\} = \left(\frac{2p}{1 + \sqrt{1 - 4p(1-p)}}\right)^m = \left(\frac{2p}{1 + |1 - 2p|}\right)^m. \tag{3.53}$$

Hence, if $p \ge \frac{1}{2}$,

$$P\{T_U < \infty\} = 1,$$

which means the geometric random walk will hit the upper barrier sooner or later. If $p < \frac{1}{2}$,

$$P\{T_U < \infty\} = \left(\frac{p}{1-p}\right)^m.$$

Thus there is a positive probability that the geometric random walk never hits the upper barrier. When $p \ge \frac{1}{2}$ the geometric random walk has an upward tendency and similarly when $p < \frac{1}{2}$ the geometric random walk has a downward tendency. Hence these results agree with intuition.

The formula (3.52) may also be used to compute the mean and higher moments of T_U. For example, Taking the first derivative of (3.52) with respect to z and letting $z = 0$ yields

$$E\{T_U \, \mathbf{I}_{\{T_U < \infty\}}\} = \frac{m}{|1 - 2p|} \left(\frac{2p}{1 + |1 - 2p|}\right)^m.$$

[8]A Laplace transform $\bar{f}(z)$ of a random variable X is said to be defective if $\bar{f}(0) < 1$. A defective Laplace transform indicates that the random variable X is not finite, i.e. $P\{|X| = \infty\} > 0$.

Thus

$$E\{\mathcal{T}_U | \mathcal{T}_U < \infty\} = \frac{E\{\mathcal{T}_U \mathbf{I}_{\{\mathcal{T}_U < \infty\}}\}}{P\{\mathcal{T}_U < \infty\}} = \frac{m}{|1 - 2p|}.$$

Another use of the formula (3.52) is to identify the probability distribution of $\mathcal{T}_U$. First, let us consider the case of $m = 1$. The Laplace transform of $\mathcal{T}_U$ is

$$E\{e^{-z\mathcal{T}_U} \mathbf{I}_{\{\mathcal{T}_U < \infty\}}\} = \frac{2p}{e^z + \sqrt{e^{2z} - 4p(1 - p)}} = \frac{2pe^{-z}}{1 + \sqrt{1 - 4p(1 - p)e^{-2z}}}.$$
$$(3.54)$$

In order to derive the distribution of $\mathcal{T}_U$ for $m = 1$, we expand

$$\frac{2pe^{-z}}{1 + \sqrt{1 - 4p(1 - p)e^{-2z}}}$$

in terms of e^{-z}. Rewrite

$$\frac{2pe^{-z}}{1 + \sqrt{1 - 4p(1 - p)e^{-2z}}} = \frac{2pe^{-z}(1 - \sqrt{1 - 4p(1 - p)e^{-2z}})}{1 - (1 - 4p(1 - p)e^{-2z})}$$

$$= \frac{1 - \sqrt{1 - 4p(1 - p)e^{-2z}}}{2(1 - p)e^{-z}}. \qquad (3.55)$$

The Taylor expansion yields

$$\sqrt{1 - 4p(1 - p)e^{-2z}}$$

$$= \sum_{n=0}^{\infty} (-1)^n \binom{1/2}{n} 4^n p^n (1 - p)^n e^{-2nz}$$

$$= \sum_{n=0}^{\infty} (-1)^n \frac{\frac{1}{2}(\frac{1}{2} - 1) \cdots (\frac{1}{2} - n + 1)}{n!} 4^n p^n (1 - p)^n e^{-2nz}$$

$$= - \sum_{n=0}^{\infty} \frac{1 \cdot 3 \cdot 5 \cdots (2n - 3)}{n!} 2^n p^n (1 - p)^n e^{-2nz}$$

$$= -2 \sum_{n=0}^{\infty} \frac{(2n - 2)!}{(n - 1)!n!} p^n (1 - p)^n e^{-2nz}. \qquad (3.56)$$

Substitution of (3.56) into (3.55) yields

$$E\{e^{-z\mathcal{T}_U} \mathbf{I}_{\{\mathcal{T}_U < \infty\}}\} = \sum_{n=1}^{\infty} \frac{(2n - 2)!}{(n - 1)!n!} p^n (1 - p)^{n-1} (e^{-z})^{2n-1}.$$

Thus, for $m = 1$,

$$P\{\mathcal{T}_U = 2n - 1\} = \frac{(2n - 2)!}{(n - 1)!n!} p^n (1 - p)^{n-1}, \ n = 1, 2, \cdots. \qquad (3.57)$$

For arbitrary integers m, from the expression of (3.52) the hitting time T_U is the sum of m independent hitting times whose probability distribution is given in (3.57). Thus, its probability distribution is the m-fold convolution of (3.57), and can be computed accordingly.

The Laplace transform and other properties for the barrier hitting time T_U and the method we employ here can be used to price so-called digital or binary options which pay a fixed amount when the underlying asset hits a pre-specified barrier(s).

$\square$

On a final note of this section, we point out that to apply the Optional Sampling Theorem, one needs to verify the boundedness condition in the theorem. If this condition is violated, equation (3.47) might not hold. To see this, consider the following example.

Example 3.11 Let $\{X(t)\}$ be the standard random walk $\{X(t)\}$ with $p = \frac{1}{2}$, which is given in (3.2). As we have seen, the standard random walk $\{X(t)\}$ is a martingale. Let $T_h = \inf\{t; X(t) = h\}$, i.e. T_h is the first time the standard random walk $\{X(t)\}$ hits the barrier $x = h$. Then T_h is a stopping time. Example 3.10 shows that $P\{T_h < \infty\} = 1$. That is, the standard random walk $\{X(t)\}$ will hit the barrier sooner or later. Thus, $E\{X(T_h)\} = h$, since $X(T_h)$ is always equal to h and $P\{T_h < \infty\} = 1$. On the other hand, $E\{X(0)\} = 0$, since $X(0)$ is a constant and equal to 0. Therefore

$$E\{X(T_h)\} = h \neq 0 = E\{X(0)\}.$$

If the result of the Optional Sampling Theorem held, then we would have $E\{X(T_h)\} = E\{X(0)\}$. We obtain two contradictory results. In fact, $X(T_h \wedge t)$ is not bounded from below, and thus the boundedness condition of the Optional Sampling Theorem is violated.

$\square$

3.6   OPTION PRICING WITH BINOMIAL MODELS

In this section, we discuss the pricing of options when the underlying stock price follows a binomial tree as described in Section 2.2. We begin with the case where there is only one trading period. Although this is the the simplest case but many important concepts and ideas can be illustrated and explained.

Pricing Options with One-Period Binomial Model

Consider an economy with only two states: the upstate and the downstate, respectively. The probability of the upstate is $0 < p < 1$ and the probability of the downstate is $1 - p$. There are two securities: one risk-free bond with interest rate R and one stock with initial price $S(0)$ and time-1 price $S(1)$ with u and d being the returns of the stock at the upstate and the at the downstate at time 1 (Figure 3.4).

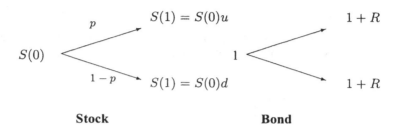

Stock   **Bond**

Figure 3.4. The returns of a stock and a bond.

A trading strategy for a portfolio is a pair (B_0, Δ_0) where B_0 is the dollar amounts in the bond and Δ_0 the number of shares of the stock at time 0. Thus, the value of the portfolio at times 0 and 1 are

$$V(0) = B_0 + \Delta_0 S(0) \text{ and } V(1) = B_0(1 + R) + \Delta_0 S(1).$$

The price system of the bond and the stock admits arbitrage if and only if there is a trading strategy (B_0, Δ_0) with $V(0) = B_0 + \Delta_0 S(0) = 0$ such that $V(1) \geq 0$ and $P\{V(1) > 0\} > 0$. It can be shown that the price system does not admit arbitrage if and only if $d < 1 + R < u$. Suppose that the condition $d < 1 + R < u$ is violated, say, $1 + R \leq d$. It is easy to verify that $\Delta_0 = 1$ and $B_0 = -S(0)$ is an arbitrage trading strategy. The case of $u \leq 1 + R$ is similar. Conversely, if $d < 1 + R < u$, then for any strategy (B_0, Δ_0) with $V(0) = B_0 + \Delta_0 S(0) = 0$,

$$V(1) = B_0(1 + R) + \Delta_0 S(1) = \begin{cases} \Delta_0 S(0)[-(1 + R) + u], \\ \Delta_0 S(0)[-(1 + R) + d], \end{cases}$$

implying that $V(1)$ can not be non-negative. In other words, there is no arbitrage strategy. In the following, we assume that the price system does not admit arbitrage and derive option prices using a no-arbitrage argument.

A (European) option is an option which gives its holder the right (not obligation) to receive or pay a pre-specified amount that is contingent on the state of the economy on a specific date. For instance, a European call option written on a risky security gives its holder the right (not obligation) to buy the underlying security at a pre-specified price on a specific date; while a European put option written on a security gives its holder the right (not obligation) to sell the underlying security at a pre-specified price on a specific date. The pre-specified price is called the strike price and the specific date is called the expiration or maturity time. The payoff functions for a European call option and a European put option are

$$\max\{S(T) - K, \ 0\}, \ \text{and} \ \max\{K - S(T), \ 0\}, \tag{3.58}$$

respectively, where K is the strike price and T is the expiration time. Figures 3.5 and 3.6 show the payoff function of a call and a put.

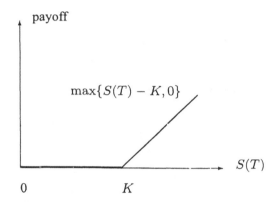

Figure 3.5. The payoff function of a call.

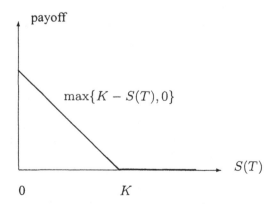

Figure 3.6. The payoff function of a put.

Consider now that there is an option which pays C_u at the upstate and C_d at the downstate at time 1. To determine the price of this option, we construct a portfolio hoping that the payoff of this portfolio is the same as that of the option. If we can do so, then the price of the option must be equal to the initial value of the portfolio in order to avoid arbitrage. For the option with payoff C_u or C_d, the corresponding portfolio is constructed as follows. The trading strategy (B_0, Δ_0) should be such that the following payoff equations are satisfied:

$$B_0(1 + R) + \Delta_0 S(0)u = C_u, \text{ at the upstate}$$
$$B_0(1 + R) + \Delta_0 S(0)d = C_d, \text{ at the downstate.}$$

Solving these equations yields

$$\Delta_0 S(0) = \frac{C_u - C_d}{u - d}, \quad B_0 = \frac{uC_d - dC_u}{(1 + R)(u - d)}. \tag{3.59}$$

Therefore, the price of the option is given by

$$B_0 + \Delta_0 S(0) = (1 + R)^{-1} [C_u q + C_d(1 - q)]. \tag{3.60}$$

where $q = \frac{1+R-d}{u-d}$, the same as (3.40) in Section 3.4. Note that $0 < q < 1$ since $d < 1 + R < u$.

Formula (3.59) suggests that to meet the obligation of this option, we invest $\frac{C_u-C_d}{u-d}$ dollars in the stock and $\frac{uC_d-dC_u}{(1+R)(u-d)}$ dollars in the bond (the sign of these values indicates whether we take a long position or a short position: a positive sign for long and a negative sign for short). The quantity $\Delta_0 = \frac{C_u-C_d}{S(0)(u-d)}$ not only represents the number of shares in the stock but also is a measure of the sensitivity of the option price to the stock price. For this reason, it is often referred to as delta [9]. The method of constructing a portfolio to perfectly hedge the payoff of an option is called delta hedging.

We can now make some very interesting observations from formula (3.60): (i) since $0 < q < 1$, $\{q, 1 - q\}$ may be viewed as a probability measure associated with the upstate and the downstate. Thus, it follows from (3.60) that the price of the option is the expected discounted payoff of the option under the probability measure $\{q, 1 - q\}$;
(ii) under the probability measure $\{q, 1 - q\}$, the stock and the bond have the same expected return which is $1 + R$;
(iii) the new probabilities q and $1 - q$ depend only on the returns of the stock and the bond, neither on the 'physical' probability measure $\{p, 1 - p\}$ nor on the payoff of the option.
In finance, the probability measure $\{q, 1 - q\}$ is referred to as the risk-neutral

[9]There are other measures referred to as Greeks. See Hull (1993), Section 14.4.

(probability) measure or Q-measure, and the probability measure $\{p, 1 - p\}$ the physical measure or P-measure.

Observation (ii) indicates that the approach used above is equivalent to one in which we change the probabilities so that the return of the stock is equal to the return of the bond, the same idea we used in the previous section where we adjusted the probabilities of the jump sizes in a random walk to meet a given drift.

Consider now the European call and put. We assume that $S(0)d < K < S(0)u$. Other cases are trivial. As in (3.58), for a call,

$$C_u = S(0)u - K \text{ and } C_d = 0.$$

Thus, $uC_d - dC_u = -d[S(0)u - K] < 0$. It follows from (3.59) that $B_0 < 0$, meaning that to hedge a call we always take a short position in bond and a long position in stock. To hedge a put we do just the opposite, by taking a long position in bond and a short position in stock, since $uC_d - dC_u = u[K - S(0)d] > 0$.

Example 3.3   Consider a stock with a current price of 100 and a riskfree bond with an interest rate of 10%. There are two possible prices for the stock: 120 or 90, at time 1. An at-the-money European call option expiring at time 1 is written on the stock[10]. That is, $K = 100$. In this case, $u = 1.2, d = 0.9, C_u = 20$ and $C_d = 0$. From formulae (3.59) and (3.60),

$$S(0)\Delta_0 = 66.67, \ B_0 = -54.55,$$

so the price of this option is

$$B_0 + \Delta_0 S(0) = 12.12.$$

In order to fulfil the option commitment, the option writer needs to be short 54.55 in bond and to be long 66.67 in stock.

It is easy to see that $q = 2/3$. Thus, the risk-neutral measure or Q-measure is $\{2/3, 1/3\}$, which can be used to price other options written on the same stock.

Suppose that we want to price an out-of-money put on the stock with $K = 95$. In this case, $C_u = 0$ and $C_d = 5$. From (3.60) the price of this option is

$$E_Q\{(1 + R)^{-1} \max(K - S(1), 0)\} = \frac{1}{1.1} \left[\frac{2}{3} \cdot 0 + \frac{1}{3} \cdot 5\right] = 1.56,$$

---

[10]An option is called at the money if no money is made by either the option holder or the option writer when the option is exercised immediately. Similarly, an option is called in the money if the immediate exercise of the option makes money and out of the money if the immediate exercise of the option loses money.

where E_Q denotes the expectation under the Q-measure. Similarly, we can use the Q-measure to evaluate the price of more complex options such as strangles[11]. A strangle involves positions in two options on the same stock. It is created by buying a put with a relatively low strike price and buying a call with a relatively high strike price. This strategy leads to a profit when the future stock price is lower than the low strike price or higher than the high strike price (Figure 3.7).

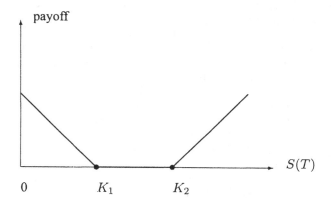

Figure 3.7. The payoff function of a strangle.

Consider a strangle with $K_1 = 95$ and $K_2 = 110$. The payoff of this strangle is

$$C_u = \max(K_1 - S(0)u, 0) + \max(S(0)u - K_2, 0) = 10$$

and

$$C_d = \max(K_1 - S(0)d, 0) + \max(S(0)d - K_2, 0) = 5.$$

Thus the cost of this strangle is

$$\frac{1}{1.1}\left[\frac{2}{3} \cdot 10 + \frac{1}{3} \cdot 5\right] = 7.58.$$

It is worth noting that we can value the call and the put separately and the cost will be the sum of the two. However, when the payoff function of an option is complex, it is easier to value the option by computing its expected

---

[11] For other option trading strategies, see Hull (1993), Chapter 8.

discounted payoff with repect to the Q-measure directly.

□

We now extend the results for the one-period model to a multi-period model. Assume again that there are a risk-free bond and a stock over T trading periods. For simplicity we assume that the time length of each trading period is 1. There are only two states of the economy for each trading period: the upstate and the downstate, given that we are at the beginning of the period. The probablities of the upstate and the downstate are p and $1 - p$, respectively. The return of the stock for each trading period is u when the upstate is attained and d when the downstate is attained. Let $S(t)$ be the price of the stock at time t. Then, as described in Section 3.2 the stock price follows a recombining binomial tree (Figure 3.2). The risk-free bond[12] earns interest at rate R for each trading period, i.e., the payoff of the bond with an initial value of 1 is $(1 + R)^t$ at time t. It is easy to see from Figure 3.2 that this multi-period binomial model can be decomposed into a collection of one-period models such that the payoffs of the price system of each one-period model are the initial prices on the two following one-period models.

A trading strategy for a portfolio is a pair of stochastic processes (B_t, Δ_t); $t = 0, 1, \cdots, T - 1$, on th economy, where B_t is the dollar amount in the bond and Δ_t is the number of shares of the stock held at time t. A trading strategy or portfolio is said to be self-financing if

$$B_{t-1}(1 + R) + \Delta_{t-1}S(t) = B_t + \Delta_t S(t), \qquad (3.61)$$

for $t = 1, 2, \cdots, T - 1$. In other words, under a self-financing trading strategy an investor invests no new money into or take no money out of the portfolio during the intermidiate trading periods. Let $V(t) = B_t + \Delta_t S(t)$ be the value of the corresponding portfolio at time t. The price system of the bond and the stock admits arbitrage if and only if there is a self-financing trading strategy $(B_t, \Delta_t); t = 0, 1, \cdots, T - 1$, with $V(0) = 0$ such that $V(T) \geq 0$ and $P\{V(T) > 0\} > 0$.

We now show that the price system does not admit arbitrage if and only if none of the one-period models admits arbitrage[13]. Suppose that there is a one-period model that admits arbitrage. We thus do nothing until the state that represents the start of the one-period model prevails. Since this one-period model admits arbitrage, we then implement an arbitrage trading strategy and invest all the proceeds into the bond at the end of the trading period. This will give us a self-financing arbitrage trading strategy. Conversely, suppose that none of the one-period models admits arbitrage but there is a self-financing arbitrage portfolio with an initial value of 0 for the multi-period model. Under

---

[12]It is commonly called the money market account as a bond can be subject to interest rate risk as we will see in the next section.

[13]This is also true for a general multi-nomial model, under which there are a finite number of states for each trading period and the length of its time steps and its volatility can vary over time.

the arbitrage strategy all the possible payoffs at T are nonnegative and at least one of them is strictly positive. No arbitrage for all the one-period models over the period $[T - 1, T]$ implies that the possible values of the portfolo at $T - 1$ are nonnegative and at least one of them is strictly positive. Continuing backwards in time in this manner, we obtain a positive initial value for the portfolio, a contradiction. An immediate implication is that the multi-period binomial model does not admit arbitrage if and only if $d < 1 + R < u$, as the latter is the no-arbitrage condition for the one-period model.

Consider now a European option that pays $C(T)$ at expiration time T. Similar to the one-period case, it can be priced using a no-arbitrage argument. That is, if there is a self-financing portfolio such that the value of the portfolio, $V(T)$, at time T is equal to $C(T)$, then the option price must be the same as the initial value $V(0)$ of the portfolio to avoid arbitrage. Moreover, the price of the option at any intermediate time t for the same reason is the value $V(t)$ of the portfolio. To identify such a self-financing portfolio, we use the delta hedging for the one-period model described earlier in this section. Starting with the time period $[T - 1, T]$, there is a delta hedging strategy at time $T - 1$ for payoff $C(T)$ as described in the one-period model and we denote it as (B_{T-1}, Δ_{T-1})[14]. Note that they are random variables at time $T - 1$. Let $V(T - 1) = B_{T-1} + \Delta_{T-1}S(T - 1)$ be the value of the corresponding portfolio at time $T - 1$. As shown in the one-period model, $V(T - 1)$ can be expressed as an expected discounted payoff

$$V(T - 1) = E_Q\{(1 + R)^{-1}C(T)|S(T - 1)\},$$

where the conditional probabilities q and $1 - q$ of the upstate and downstate are given in (3.40) of Section 3.4. Next, we move over to the time period $[T - 2, T - 1]$ and let $V(T - 1)$ be the payoff at the end of the period. Again, a delta hedging startegy (B_{T-2}, Δ_{T-2}) can similarly be identified and

$$V(T - 2) = E_Q\{(1 + R)^{-1}V(T - 1)|S(T - 2)\},$$

with the same conditional upstate and downstate probabilities q and $1 - q$. The delta hedging startegy (B_{T-2}, Δ_{T-2}) implies

$$B_{T-2}(1 + R) + \Delta_{T-2}S(T - 1) = B_{T-1} + \Delta_{T-1}S(T - 1),$$

a self-financing condition at time $T - 1$.

In general, we have a self-financing trading strategy (B_t, Δ_t) such that $V(t) = B_t + \Delta_t S(t)$, $t = 0, 1, \cdots, T - 1$ and

$$V(t) = E_Q\{(1 + R)^{-1}V(t + 1)|S(t)\}, \tag{3.62}$$

with $V(T) = C(T)$. The conditional upstate and downstate probabilities for each period remain the same and they are q and $1 - q$ given in (3.40). Thus, we

---

[14]More precisely, for each state at $T - 1$, there is a pair of values representing the delta hedging and all of them together form the delta hedging startegy (B_{T-1}, Δ_{T-1}).

rederive the martingale probability measure determined by (3.38) and (3.40). Furthermore, the price of the option is given by

$$\phi_C = V(0) = E_Q \left\{ (1 + R)^{-T} C(T) \right\}, \tag{3.63}$$

which is due to the Law of Iterated Expectation. Recall that the discounted price process $S^*(t) = (1 + R)^{-t} S(t)$ is a martingale. That implies that

$$E_Q \{S(T)\} = (1 + R)^T,$$

i.e. $S(t)$ has the same return as that of the risk-free bond under the probability measure Q. Hence the probability measure Q is called not only the Q-measure (for the obvious reason) or the martingale measure but also the risk-neutral measure. The valuation of an option based on (3.63) is often referred to as the risk-neutral valuation.

Fundamental Theorem of Asset Pricing
Formula (3.63) can be extended to a very general situation. Suppose that a market has a finite number of primitive tradable risky securities $S_i(t), i = 1, \cdots, n$, and a money market account $B(t), B(0) = 1$, which earns interest at instantaneous rate $r(t)$ over time period $[0, T]$. If there exists a martingale measusre such that $B(t)^{-1} S_i(t), i = 1, \cdots, n$, are martingales, then the price system of these securities does not admit arbitrage[15] and any option[16] expiring at T can be priced as

$$\Phi_C = E_Q \left\{ B(T)^{-1} C(T) \right\}. \tag{3.64}$$

See Harrison and Pliska (1981).

We now apply (3.63) to derive a closed form expression for the price of a European call. Let $C(T) = \max(S(T) - K, 0)$ be the payoff of a call with strick price K. It follows from (3.63) that the price of the call is

$$\Phi_C = E_Q\{(1 + R)^{-T} C(T)\}$$

$$= (1 + R)^{-T} \sum_{s=0}^{T} \max(S(0) u^s d^{T-s} - K, 0) \begin{pmatrix} T \\ s \end{pmatrix} q^s (1 - q)^{T-s}$$

$$= (1 + R)^{-T} \sum_{s \geq \frac{\log(K/S(0)) - T \log d}{\log(u/d)}} (S(0) u^s d^{T-s} - K) \begin{pmatrix} T \\ s \end{pmatrix} q^s (1 - q)^{T-s}$$

$$= (1 + R)^{-T} S(0) \sum_{s \geq \frac{\log(K/S(0)) - T \log d}{\log(u/d)}} \begin{pmatrix} T \\ s \end{pmatrix} u^s d^{T-s} q^s (1 - q)^{T-s}$$

---

[15]Its reverse is true under some technical conditions but for pricing purposes we only need to find a martingale measure and apply this half of the theorem.
[16]It is more suitable to call it a contingent claim but we use the term option for simplicity.

$$- \quad (1+R)^{-T}K \sum_{s \geq \frac{\log(K/S(0)) - T\log d}{\log(u/d)}} \binom{T}{s} q^s (1-q)^{T-s}.$$

Let

$$\bar{q} = \frac{uq}{1+R}. \tag{3.65}$$

The price of the call can then be written as

$$
\begin{aligned}
\Phi_C \quad = \quad & S(0) \sum_{s \geq \frac{\log(K/S(0)) - T\log d}{\log(u/d)}} \binom{T}{s} \bar{q}^s (1-\bar{q})^{T-s} \\
& - \quad (1+R)^{-T}K \sum_{s \geq \frac{\log(K/S(0)) - T\log d}{\log(u/d)}} \binom{T}{s} q^s (1-q)^{T-s} \\
= \quad & S(0) \sum_{s \leq \frac{T\log u + \log(S(0)/K)}{\log(u/d)}} \binom{T}{s} (1-\bar{q})^s \bar{q}^{T-s} \\
& - \quad (1+R)^{-T}K \sum_{s \leq \frac{T\log u + \log(S(0)/K)}{\log(u/d)}} \binom{T}{s} (1-q)^s q^{T-s}.
\end{aligned}
$$

$$\tag{3.66}$$

Denote $x_0 = \frac{T\log u + \log(S(0)/K)}{\log(u/d)}$ and $B(x;\, n,p)$ the distribution function of the binomial distribution with parameters n and p, i.e.

$$B(x;\, n,p) = \sum_{s \leq x} \binom{n}{s} p^s (1-p)^{n-s}. \tag{3.67}$$

We have

$$\Phi_C = S(0)B(x_0;\, T, 1-\bar{q}) - (1+R)^{-T}KB(x_0;\, T, 1-q). \tag{3.68}$$

This formula is the well known option pricing formula of Cox, Ross and Rubinstein (1979). It also shows the delta hedging strategy of the call under the binomial model.

To price a European put option, we can either use the above approach or use the so-called put-call parity. Let $P(T) = \max(K - S(T),\, 0)$ be the payoff of a European put with strike price K. It is easy to see that

$$\max(S(T) - K,\, 0) - \max(K - S(T),\, 0) = S(T) - K.$$

Hence if Φ_P is the price of the put, it follows from (3.63) that

$$\Phi_C - \Phi_P = E_Q\left\{ (1+R)^{-T} \left(S(T) \right) - K \right\} = S(0) - (1+R)^{-T}K. \tag{3.69}$$

This identity is called the put-call parity.

Finally in this section, we discuss how to price American type of options. The payoff structure of an American option is similar to its European counterpart, except that it can be exercised anytime before or on its expiration date. For example, an American call option and an American put option written on a stock with price $S(t)$ at time t for the period $[0, T]$ can be exercised before time T. Their payoffs, if exercised at t, will be $\max(S(t) - K, 0)$ and $\max(K - S(t), 0)$, respectively.

The valuation of American options is generally much more difficult than European options. Most American options do not have a closed form solution for their prices. This is because the holder of an American option has the right to exercise anytime and the valuation problem in turn becomes how to find the optimal exercise time at which the expected discounted payoff under a risk-neutral probability measure is maximized. However, if a binomial model is used for the stock price, the price of an American option can be computed recursively.

Let $g(S(t), t)$ be the payoff of an American option when it is exercised at time t. Suppose that T is a time the holder of the option wants to exercise it. It is natural to assume that the holder's decision to exercise or not is based on the information up to date. Thus, T is a stopping time introduced in Section 3.5. In this case, it follows from formula (3.63) or (3.64) that the price of the American option with exercise time T is given by[17]

$$E_Q\{(1 + R)^{-T} g(S(T), T)\}. \tag{3.70}$$

Since the holder of the option, not the writer, has the right to exercise and has an opportunity to maximize the payoff, the option price hence is

$$\Phi_g = \max_T E_Q\{(1 + R)^{-T} g(S(T), T)\}. \tag{3.71}$$

Maximization is taken over all possible stopping times over the period $[0, T]$. It is easy to see that there is no put-call parity for American options since the optimal exercise time for a call is different from the optimal exercise time for the corresponding put.

It is impractical to examine each of these stopping times in (3.71) in order to find the optimal exercise time and the price of the option. However, under the discrete-time framework we will be able to find the optimal exercise time and the option price through a backward recursive algorithm.

We begin with the last time period $[T - 1, T]$. Define a random variable $V(T - 1, \mathcal{F}_{T-1})$ on $(\Omega, \mathcal{F}_{T-1})$ as

$$V(T - 1, \mathcal{F}_{T-1})$$

---

[17] Readers may notice that formula (3.70) is slightly different from (3.63) as the option payoff is discounted at the exercise time not the expiration time. However, they are equivalent if we assume that the option payoff is re-invested in the money market account at the exercise time.

$$= \max\{(1+R)^{-1}E_Q[g(S(T),T)|\mathcal{F}_{T-1}],\ g(S(T-1),T-1)\}. \tag{3.72}$$

For $t = 1, \cdots, T-1$, define a random variable $V(t-1, \mathcal{F}_{t-1})$ on $(\Omega, \mathcal{F}_{t-1})$ as

$$V(t-1, \mathcal{F}_{t-1})$$
$$= \max\{(1+R)^{-1}E_Q[V(t, \mathcal{F}_t)|\mathcal{F}_{t-1}],\ g(S(t-1), t-1)\}. \tag{3.73}$$

The value $V(0)$ then is the price of this American option at time 0. Furthermore, $V(t, \mathcal{F}_t)$ is the price of the option at time t. In other words, the price of an American option is calculated as the maximum of the expected discounted value of the same option at next trading date and the current payoff. The optimal exercise time of this option then is the stopping time

$$\mathcal{T}_g = \min\{t; g(S(t), t) > V(t, \mathcal{F}_t)\}, \tag{3.74}$$

where if the set is empty, we define $\mathcal{T}_g = T$.

The rationale behind this algorithm is simple: choose $\mathcal{T}_0 = T$ as an initial exercise time which of course is not optimal. If at time $T-1$, the stock price is $S(T-1) = s_1$ and

$$g(s_1, T-1) > (1+R)^{-1}E_Q\{g(S(T), T)|S(T-1) = s_1\},$$

we define

$$\mathcal{T}_1 = \begin{cases} T-1, & \text{at } S(T-1) = s_1 \\ \mathcal{T}_0, & \text{elsewhere.} \end{cases}$$

Thus the exercise time $\mathcal{T}_1$ will yield the same or a higher expected discounted payoff than $\mathcal{T}_0$. That is, $\mathcal{T}_1$ is a better exercise time. Choose another value of $S(T-1)$, say $S(T-1) = s_2 \neq s_1$ and repeat the same step. If

$$g(s_2, T-1) > (1+R)^{-1}E_Q\{g(S(T), T)|S(T-1) = s_2\},$$

define

$$\mathcal{T}_2 = \begin{cases} T-1, & \text{at } S(T-1) = s_2 \\ \mathcal{T}_1, & \text{elsewhere.} \end{cases}$$

Hence $\mathcal{T}_2$ is better than $\mathcal{T}_1$. Going through all the possible values of $S(T-1)$ (only a finite number of them), we obtain the optimal exercise time for the period $[T-1, T]$. Moreover, the value of the option at time $T-1$ is $V(T-1, \mathcal{F}_{T-1})$. Repeating the same procedure for $[T-2, T-1]$ and the rest of intermediate trading periods, we then obtain the optimal exercise time and price of the option. The algorithm we discussed here is quite flexible. It can apply to other types of options. For instance, we may use it to evaluate Bermudan options which allow their holders to exercise during a given period of time before expiration of the options. In that case, we may use the algorithm for the exercise period and use an option pricing formula for European options for the no exercise period.

We end this section by showing that it is optimal not to exercise an American call option early if its underlying stock pays no dividend. An immediate implication is that its price is the same as that of the corresponding European call option. It is sufficient to show that for any t,

$$\max\{S(t-1) - K,\, 0\} \le (1 + R)^{-1} E_Q\{\max\{S(t) - K,\, 0\}|S(t-1)\}.$$
$$(3.75)$$

The inequality (3.75) is in fact a direct application of Jensen's Inequality which states that for any random variable X and for any concave up function $h(x)$, $h\{E(X)\} \le E\{h(X)\}$. Now choose $h(S) = \max(S - K,\, 0)$. We have

$$
\begin{aligned}
&(1 + R)^{-1} E_Q\{\max\{S(t) - K,\, 0\}|S(t-1)\} \\
\ge\ &(1 + R)^{-1} \max\{E_Q\{S(t)|S(t-1)\} - K,\, 0\} \\
=\ &\max\{S(t-1) - \frac{K}{1+R},\, 0\} \ge \max\{S(t-1) - K,\, 0\}.
\end{aligned}
$$

3.7 BINOMIAL INTEREST RATE MODELS

In this section, we illustrat how a binomial model can be used to model interest rates. Two interest rate models will be considered. The first model is the short rate model of Black, Derman and Toy (1990), which has been widely used in practice. The second model is the Ho-Lee forward rate model (Ho and Lee, 1986), which may be viewed as a one-factor HJM model (Heath, Jarrow and Morton, 1990).

We begin with the introduction of four types of interest rates in a discrete-time setting[18]. The first interest rate is the t-year zero-coupon interest rate. This is the rate used to calculate interest payments on a zero-coupon bond that will mature in t years. The t-year zero-coupon interest rate is often referred to as the t-year zero rate or t-year spot rate. The second interest rate is the yield to maturity rate or simply the yield to maturity or the bond yield. This is the annual internal rate of return on a coupon bond. Since we may view a zero-coupon bond as a coupon bond with zero coupon payment, the zero rate is the yield to maturity rate for a zero-coupon bond, and for this reason we use both terms exchangeably for zero-coupon bonds. The remaining two interest rates are the forward rate and the short interest rate (also called short rate) which we will describe later.

---

[18]The continuous version will be introduced and discussed in Chapter 5, Section 3.

We begin with zero rates. Consider a bond market in which default-free zero-coupon bonds[19] are traded during the period $[0, T]$. For simplicity we assume that the trading time is measured in integer units $0, 1, 2, \cdots, T$, i.e. $\tau = 1$. Let $P(t, s)$ be the price of a default-free zero-coupon bond at time t which pays one monetary unit at maturity time s.

The $(s - t)$-year zero rate at time t is defined as

$$y(t,s) = \left[\frac{1}{P(t,s)} \right]^{\frac{1}{s-t}} - 1. \qquad (3.76)$$

That is,

$$P(t, s) = [1 + y(t, s)]^{-(s-t)}. \qquad (3.77)$$

Thus, the zero rate[20] $y(t, s)$ is the rate of return, compounding annually, on a zero-coupon bond over the period $[t, s]$. A zero-coupon yield curve, or zero curve is a smooth curve that plots the zero rate versus the time to maturity for zero-coupon bonds maturing in the future. The zero curve is also referred to as the spot curve by many practitioners.

As mentioned earlier, the yield to maturity rates are the rates of return investors earn when purchasing default-free coupon bonds. In the U.S. bond market, they are represented by the U.S. treasury securities[21]. Quotations on the U.S. treasury securities such as prices and yields are publically available. The table in the next page is a sample of quotes from a financial publication.

In the following, we use a U.S. treasury security to illustrate how a yield to maturity rate is calculated. We assume that this treasury security has a maturity greater than one year so it is a coupon bond with semiannual coupon payments. Suppose that the treasury has a par value of F dollars and the annual coupon rate of c, and it matures in n years. The current market price of the treasury is P. Let $C = cF/2$. Then C is the amount of each semiannual coupon payment for the bond. The yield to maturity y on the bond, convertible semiannually, is calculated using the following formula:

$$F(0) = \frac{C}{(1 + 0.5y)} + \frac{C}{(1 + 0.5y)^2} + \cdots + \frac{C}{(1 + 0.5y)^{(2n-1)}} + \frac{C + F}{(1 + 0.5y)^{2n}}, \qquad (3.78)$$

where there are $2n$ terms on the right-hand side since that is the total number of payments of the bond till maturity. Clearly, the yield to maturity y is the

---

[19]Zero-coupon bonds are also called discount bonds or strip bonds.

[20]An alternative definition is to assume that yield rates are compounded continuously. In this case, $y(t, s) = -\frac{1}{s-t} \ln P(t, s)$. Thus, $P(t, s) = e^{-y(t,s)(s-t)}$. The advantage of this definition is that it provides a unified treatment for both discrete and continuous models.

[21]U.S. Treasury securities with maturities of one year or less are called Treasury bills. Bills are in fact discount bonds which pay only a contractual fixed amount at maturity. Bills are quoted on a bank discount basis and are based on a 360-day year. Treasury securities with maturities between 2 and 10 years are called Treasury notes and those with maturities greater than 10 years are called Treasury bonds. Notes and bonds pay coupons every six months plus principal at maturity. See Fabozzi (1997), Chapter 8 for details.

Bills	Coupon	Maturity Date	Price(Yield)
3 month		9/30/99	4.63(4.76)
6 month		12/30/99	4.83(5.03)
1 year		6/22/00	4.79(5.05)
Notes/Bonds			
2 year	5.75	6/30/01	100 12/32(5.54)
5 year	5.25	5/15/04	98 10/32(5.65)
10 year	5.50	5/15/09	97 27/32(5.79)
30 year	5.25	2/15/29	90 02/32(5.97)

Table 3.1.   A sample of quotes on U.S. Treasuries.

solution of the nonlinear equation (3.78), and hence needs to be solved using a numerical method. To solve the equation, we may use the Newton-Raphson method or the bisection method, for example. Similar to the zero curve, the bond yield curve is a graphical plot of the yields of the coupon bonds versus their maturities. Yield curves of Treasury securities available daily in the *Wall Street Journal* and other financial publications are the yield curves of this type since the Treasury securities with maturity greater than one year are coupon bonds. Those yield curves are constructed using the market data. Although the bond yield curve is readily available, it is not very useful when we want to price a bond that has the same quality but a different payment pattern from the Treasury securities. On the other hand, any bond can be priced easily using the zero curve because each payment of the bond can be viewed as a zero-coupon bond maturing at the time of payment. Hence, it is necessary to derive the zero curve from the bond yield curve. It can be shown that there is a one-to-one correspondence between the zero curve and the bond yield curve, which makes the derivation of the zero curve possible[22].

A forward rate is an interest rate both parties of a financial transaction agree today to use for calculating payments in a future period of time. In other words, it is a future zero rate that is determined now. Let $f(t, s)$ be the forward rate at time t for the future period $[s, s+1]$, $s \geq t$, i.e. the interest rate both parties agree to at time t for the period $[s, s+1]$. To calculate $f(t, s)$, we consider the following transaction: paying one dollar at time s and receiving a fixed amount at time $s + 1$. Since $f(t, s)$ is the rate for the period $[s, s+1]$, the amount received at $s+1$ must be $[1 + f(t, s)]$. Note that the time-t value of one dollar at s is $P(t, s)$ and the time-t value of $[1 + f(t, s)]$ dollars at $s+1$ is $[1 + f(t, s)]P(t, s + 1)$. Since there is no money changing hands at time t

---

[22] To calculate a zero rate for a given time period, we need to know not only the price of a coupon bond over the same time period but the coupon bond maturing at each coupon payment date of the first coupon bond. Somtimes in practice, we do not have such information and in that case, a linear interpolation or other approximation technique is used. This is illustrated in Figure 3.8.

under this agreement, we must have

$$P(t, s) = [1 + f(t, s)]P(t, s + 1).$$

Thus

$$f(t, s) = \frac{P(t, s)}{P(t, s + 1)} - 1, \qquad (3.79)$$

or

$$P(t, s + 1) = P(t, s) \left[\frac{1}{1 + f(t, s)} \right], \qquad (3.80)$$

for $t = 0, 1, 2, \cdots, T$; $s = t, \cdots, T$. The formula (3.80) relates two default-free zero-coupon bonds with different maturities. Applying (3.80) recursively yields

$$P(t, s) = \left[\frac{1}{1 + f(t, t)} \right] \left[\frac{1}{1 + f(t, t + 1)} \right] \cdots \left[\frac{1}{1 + f(t, s - 1)} \right]. \qquad (3.81)$$

Hence a bond price $P(t, s)$, for any $t \leq s$, is uniquely determined by forward rates. Further, comparing (3.77) with (3.81), we have

$$[1 + y(t, s)]^{(s-t)} = [1 + f(t, t)][1 + f(t, t + 1)] \cdots [1 + f(t, s - 1)], \qquad (3.82)$$

for all $t \leq s$. It follows from (3.82) that

$$[1 + y(t, s + 1)]^{(s-t+1)} = [1 + y(t, s)]^{(s-t)}[1 + f(t, s)].$$

Thus

$$f(t, s) = \frac{[1 + y(t, s + 1)]^{(s-t+1)}}{[1 + y(t, s)]^{(s-t)}} - 1, \qquad (3.83)$$

and in particular $f(t, t) = y(t, t + 1)$. The formula (3.82) or (3.83) displays a one-to-one correspondence between the term structure of the zero rates and the term structure of the forward rates. This allows for the derivation of the forward rate curve from the zero curve. Recalling from the discussion earlier that we may produce the zero curve using the bond market data via the coupon bond yield curve, we are able to produce the forward rate curve from the market data too. Figure 3.8 in the next page shows the relationship between the Treasury yield curve, the corresponding Treasury zero curve, and the Treasury forward rate curve, using the quotations in Table 3.1. The numerical algorithm to construct these curves involves linear interpolation technique similar to that given in Hull (1993), Chapter 4.

The last, and probably the most useful, type of interest rates for modelling purposes is short interest rates, or simply, short rates. A short rate $R(t)$ at time t is the zero rate for the immediate trading period $[t, t + 1]$. Thus, $R(t) = y(t, t + 1)$, or $R(t) = f(t, t)$ following from (3.83). Introducing

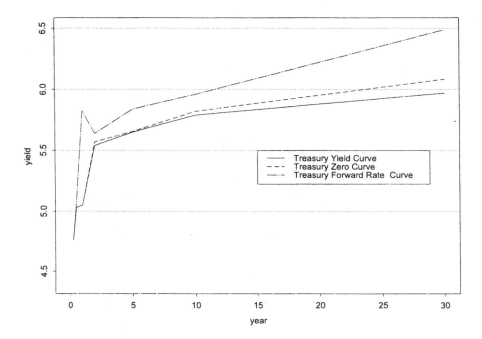

Figure 3.8.   Treasury yield curve, Treasury zero curve, and Treasury forward rate curve based on the quotes in Table 3.1.

the short rates might look redundant as they are zero rates or forward rates. However, it is necessary to separate the short rates from other zero rates as the short rates play a special role in charaterizing the dynamic of bond prices over time. Since

$$R(t) = y(t, t + 1) = \frac{1}{P(t, t + 1)} - 1,$$

we have

$$P(t, t + 1) = \frac{1}{1 + R(t)}.$$

That is, the short rate $R(t)$ is the discount rate for the time period $[t, \ t + 1]$. For a default-free zero-coupon bond with the par value of one monetary unit, maturing at time s, the time-t discounted value[23] is given by

$$D(t, s) = \left[\frac{1}{1 + R(t)}\right]\left[\frac{1}{1 + R(t + 1)}\right]\cdots\left[\frac{1}{1 + R(s - 1)}\right]. \quad (3.84)$$

Thus the time-t price of the bond, following from the Fundamental Theorem of Asset Pricing (3.64), may be expressed as

$$P(t, s) = \left[\frac{1}{1 + y(t, s)}\right]^{(s-t)}$$

$$= E_Q\left\{\left[\frac{1}{1 + R(t)}\right]\left[\frac{1}{1 + R(t + 1)}\right]\cdots\left[\frac{1}{1 + R(s - 1)}\right]\Bigg| \mathcal{F}_t\right\},$$
$$(3.85)$$

where the expectation is taken under the risk-neutral probability measure. The formula (3.85) gives a relationship between the short rate process $\{R(t),\ t = 0, 1, 2, \cdots\}$ and bond prices. If we could correctly model the short rate process $\{R(t),\ t = 0, 1, 2, \cdots\}$, we would be able to price these bonds.

As we have seen above, the price of a default-free zero-coupon bond is completely determined by the term structure of zero rates, the term structure of forward rates, or the term structure of short rates. Hence, modelling bond prices is equivalent to modelling one of these term structures. However, in practice we tend not to model the term structure of zero rates for the reason that different zero rates have different times to maturity. As a result, the inter-relation between the zero rates is difficult to analyze. This is not the case for forward rates and short rates as each forward rate or short rate is an interest rate for one trading period. A model involving the term structure of forward rates is called a forward rate model. In this case, for each fixed s the forward rate $f(t, s)$ is assumed to follow a stochastic process in t. The formula (3.81) then provides a pricing formula for a default-free zero-coupon bond. An advantage to using a forward rate model is that it often has the flexibility to produce the same zero rate curve as that from the bond market. However, since the formula (3.81) involves forward rates with different maturities, the modelling process often becomes complicated so that it is less tractable mathematically and numerically. The second approach is to model the term structure of short rates. A model of this nature is called a short rate model. In this case, it is assumed that the short rates $R(t), t = 0, 1, 2, \cdots$, follow a stochastic process. Hence a short rate model involves only one underlying stochastic process, which makes the implementation of the model easier than a forward rate model. Of course, there are some trade-offs: a short rate model might not reproduce the market zero curve and/or bond prices could be too

---

[23] The discounted value of a future payment is different from the price of the payment. The former is the equivalent value of the payment and is hence a random variable at time of payment. The latter is a fixed amount at the evaluation time in exchange of the payment.

sensitive to the model parameters. Nevertheless, a short rate model is often a preferred choice for many financial engineers.

In the previous application, we showed how to model a stock price over a period of time. We have assumed that the drift and volatility of the stock remain constant during the entire period. This restriction often needs to be modified in order to meet market conditions. This is particularly true when modelling interest rates for the bond market. Bonds at different maturities tend to have different yields, and their values and the corresponding interest rates tend to have time-dependent volatilities. In this application, we extend the binomial approach to the case where the volatility varies over time and apply it to modelling interest rates.

The Short Rate Model of Black, Derman and Toy

In the following, we present a widely used short rate model that was proposed by Black, Derman and Toy (1990). A detailed analysis can also be found in Panjer (1998), Chapter 7.

As before, let $R(t)$ be the short rate at time t for $t = 0, 1, 2, \cdots$. Suppose that the short rate process $\{R(t)\}$ follows a geometric random walk, i.e.

$$R(t) = R(0)e^{Y_1 + Y_2 + \cdots + Y_t} = R(t-1)e^{Y_t}. \tag{3.86}$$

The volatility $\sigma(t)$ of $\{R(t)\}$ at time t is defined as

$$\sigma^2(t) = Var_{(t-1)}\left\{\ln\left(\frac{R(t)}{R(t-1)}\right)\right\}, \tag{3.87}$$

where $Var_t\{\}$ denotes the conditional variance with repect to the time-t information structure. It is assumed that the volatilities are deterministic but vary over time. Further assume that the risk-neutral probablities of the upstate and the downstate are equal, i.e., $q = 1 - q = 1/2$. Let the returns of the upstate and the downstate at time t be u_t and d_t, i.e. at the upstate $e^{Y_t} = u_t$, and at the downstate $e^{Y_t} = d_t$. Then, it follows from (3.86) and (3.87) that

$$\sigma(t) = \sqrt{Var(Y_t)} = \frac{1}{2}\ln\frac{u_t}{d_t}, \tag{3.88}$$

since $Var(Y_t) = \frac{1}{4}\left(\ln\frac{u_t}{d_t}\right)^2$. Notice that d_t and u_t are usually functions of time t. Thus, the resulting binomial tree is not recombining and hence is not favorable in many financial applications. To overcome this shortcoming, we adjust the size of the up-jump and the down-jump at each node to force the binomial tree to be recombining. This can be done as follows: at time 1, we choose the random return as usual to match volatility $\sigma(1)$; at time 2, we begin with the lower node at time 1 and construct a one-period subtree with volatility $\sigma(2)$. We then construct a one-period subtree for the upper node at time 1 such that the lower node of this new subtree is the upper node of the first subtree. The upper node of the new subtree is determined by $\sigma(2)$; in general, we begin with the lowest node at the end of a completed tree and build up subtrees in an ascending order such that each subtree has the same given volatility and the upper node of a preceeding subtree is the lower node of the following subtree. One may notice that (3.88) involves only the ratio of the jump sizes at each node. Hence the value of the short rate at the lowest node at each time is not used in the procedure and therefore can be assigned to meet additional market conditions such as the zero curve.

We now illustrate this idea in the following.

Example 3.5 Suppose that we want to model 1-year short rates over a 5-year period. We are given a market term structure[24] in Table 3.2.

Maturity t (years)	1	2	3	4	5
Zero Rate $y(0, t)$	10%	11%	12%	12.5%	13%
Zero Rate Volatility $\sigma_y(t)$	20%	19%	18%	17%	16%
Short Rate Volatility $\sigma(t-1)$		19.0%	17.2%	15.3%	13.5%

Table 3.2.   The market term structure.

We first construct a one-period tree at time 0. Let $R_u(1)$ and $R_d(1)$ be the possible short rates for the second year. Then by (3.88),

$$\sigma(1) = \frac{1}{2} \ln \frac{R_u(1)}{R_d(1)} = 19\%.$$

Thus $R_u(1) = e^{2 \times 0.19} R_d(1) = 1.462 R_d(1)$. To determine $R_d(1)$, we use the formula (3.85) with $t = 0$ and $s = 2$. Thus

$$\left(\frac{1}{1 + 0.11} \right)^2 = \frac{1}{2} \left(\frac{1}{1 + R_u(1)} + \frac{1}{1 + R_d(1)} \right) \left(\frac{1}{1 + R(0)} \right).$$

The coefficient $1/2$ is the probability of the upstate and the downstate. Solving this equation for $R_d(1)$ gives $R_d(1) = 9.79\%$ and $R_u(1) = 14.32\%$ (Figure 3.9).

---

[24] The parameter values are taken from Black, Derman and Toy (1990). The original volatility values are for zero rate volatilities. In order to avoid a lengthy derivation we begin with short rate volatilities. The relationship between the zero rate volatility and short rate volatility is discussed in Panjer, et al., Chapter 7, Section 7.7.

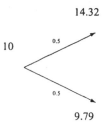

14.32

Figure 3.9. Constructing a short rate tree: step one.

We now construct the lower subtree at time 1. Let $R_u(2)$, $R_m(2)$ and $R_d(2)$ be the possible short rates for the third year. Then,

$$\sigma(2) = \frac{1}{2} \ln \frac{R_m(2)}{R_d(2)} = 17.2\%,$$

and

$$\frac{1}{2} \ln \frac{R_u(2)}{R_m(2)} = 17.2\%.$$

The above equations give that $R_m(2) = e^{2 \times 0.172} R_d(2) = 1.411 R_d(2)$ and $R_u(2) = 1.411 R_m(2) = 1.990 R_d(2)$ (Figure 3.10, the left tree). Matching the three-year bond price using (3.85) yields

$$\left(\frac{1}{1+0.12}\right)^3 = \frac{1}{4} \left[\left(\frac{1}{1+R_u(2)}\right)\left(\frac{1}{1+R_u(1)}\right)\right.$$
$$+ \left(\frac{1}{1+R_m(2)}\right)\left(\frac{1}{1+R_u(1)}\right) + \left(\frac{1}{1+R_m(2)}\right)\left(\frac{1}{1+R_d(1)}\right)$$
$$+ \left.\left(\frac{1}{1+R_d(2)}\right)\left(\frac{1}{1+R_d(1)}\right)\right]\left(\frac{1}{1+R(0)}\right).$$

By solving this equation for $R_d(2)$, we have $R_d(2) = 9.76\%$. Thus, $R_m(2) = 13.77\%$ and $R_u(2) = 19.42\%$ (Figure 3.10, the right tree, next page).

Continuing in this way, we obtain the short rate process. The complete tree of the short rates over 5 years is given in Figure 3.11. A systematic implementation of the Black-Derman-Toy model is given in Panjer (1998), Section 7.7.

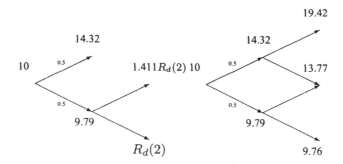

Figure 3.10. Constructing a short rate tree: step two.

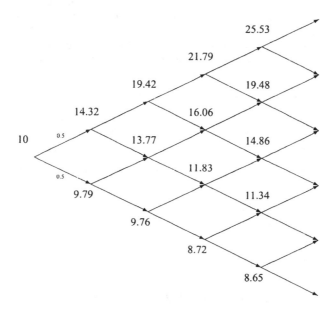

Figure 3.11. The complete short rate tree.

The Ho-Lee Forward Rate Model

As we have mentioned earlier, instead of modelling the short rate process, we may model forward rates using a binomial model. In the following, we describe a forward rate model developed by Ho and Lee (1986).

For a given time period $[0, T]$, let

$$f(t, s), \quad t = 0, 1, 2 \cdots, s,$$

be the forward rate process for time period $[s, s+1]$. We assume that it follows a random walk: for $s = 1, 2, \cdots, T - 1$, and $t = 1, 2, \cdots, s$,

$$1 + f(t, s) = [1 + f(t - 1, s)][J(t, s)]^{-1}, \tag{3.89}$$

where $J(t, s)$'s are independent Bernoulli random variables (i.e. binomial random variable with $n = 1$) such that $u(t, s) > d(t, s) > 0$ and

$$
\begin{aligned}
P_Q\{J(t, s) &= u(t, s)\} = q(t), \\
P_Q\{J(t, s) &= d(t, s)\} = 1 - q(t), \quad 0 < q(t) < 1,
\end{aligned}
\tag{3.90}
$$

where $q(t)$, $t = 1, 2 \cdots, T - 1$ are the risk-neutral probabilities[25]. The choice of $u(t, s), d(t, s)$ and $q(t)$ must be such that the bonds are priced to avoid arbitrage. In other words, their values are such that the present value processes of all the bonds are martingales. Let $A(t, s)$ be the present value process of the zero-coupon bond of 1 maturing at time s, i.e., $A(t, s) = D(0, t)P(t, s)$, where $D(0, t)$ is given in (3.84). It follows from (3.84) and (3.81) that

$$
\begin{aligned}
A(t, s) &= D(0, t)P(t, s) \\
&= D(0, t - 1)[1 + R(t - 1)]^{-1} \prod_{k=t}^{s-1}[1 + f(t, k)]^{-1} \\
&= D(0, t - 1)[1 + f(t - 1, t - 1)]^{-1} \prod_{k=t}^{s-1}[1 + f(t - 1, k)]^{-1} J(t, k) \\
&= D(0, t - 1) \prod_{k=t-1}^{s-1}[1 + f(t - 1, k)]^{-1} \prod_{k=t}^{s-1} J(t, k) \\
&= A(t - 1, s) \prod_{k=t}^{s-1} J(t, k).
\end{aligned}
\tag{3.91}
$$

Thus, $A(t, s), t = 0, 1, \cdots, s;\ s = 1, 2, \cdots, T$ are martingales if and only if

$$E_Q \left\{ \prod_{k=t}^{s-1} J(t, k) \right\} = 1. \tag{3.92}$$

---

[25] The way we present the model is a little awkward because we have not introduced continuously compounding forward rates in this section.

Or equivalently,

$$q(t) \prod_{k=t}^{s-1} u(t, k) + [1 - q(t)] \prod_{k=t}^{s-1} d(t, k) = 1. \qquad (3.93)$$

It is obvious that there are more than one choice for $u(t, k)$, $d(t, k)$ and $q(t)$ to satisfy (3.93) but to have desirable properties such as recombining for the model, we impose the following extra conditions:

- $q(t) = q$, independent of t;

- $d(t, s) = \psi u(t, s)$ for some constant $0 < \psi < 1$. This condition implies that the variance of $\ln J(t, s)$ is constant.

From (3.93), we have

$$q \prod_{k=t}^{s-1} u(t, k) + (1 - q)\psi^{s-t} \prod_{k=t}^{s-1} u(t, k) = 1,$$

or

$$\prod_{k=t}^{s-1} u(t, k) = [q + (1 - q)\psi^{s-t}]^{-1},$$

for $s = t, t + 1, \cdots, T$. We therefore obtain

$$u(t, s) = \frac{q + (1 - q)\psi^{(s-t)}}{q + (1 - q)\psi^{(s-t+1)}}, \qquad (3.94)$$

$$d(t, s) = \psi \frac{q + (1 - q)\psi^{(s-t)}}{q + (1 - q)\psi^{(s-t+1)}}. \qquad (3.95)$$

We now show that the Ho-lee model is recombining in terms of bond prices. To see this, consider a bond with maturity time s over time period $[t - 1, t + 1]$ and two possibilites: for $k = t, \cdots, s - 1$, (i) the forward rates $f(t - 1, k)$ go up first with $u(t, k)$ and then go down with $d(t + 1, k)$, and (ii) the forward rates $f(t - 1, k)$ first go down with $d(t, k)$ and then go up with $u(t + 1, k)$. In the first case,

$$
\begin{aligned}
P(t + 1, s) &= \prod_{k=t+1}^{s-1} [1 + f(t + 1, k)]^{-1} \\
&= \prod_{k=t+1}^{s-1} [1 + f(t, k)]^{-1} d(t + 1, k) \\
&= \psi^{s-t-1} \prod_{k=t+1}^{s-1} [1 + f(t, k)]^{-1} \prod_{k=t+1}^{s-1} u(t + 1, k)
\end{aligned}
$$

$$= \psi^{s-t-1} \prod_{k=t+1}^{s-1} [1 + f(t-1,k)]^{-1} \prod_{k=t+1}^{s-1} u(t,k)u(t+1,k).$$

Similarly, in the second case, we have

$$P(t+1,s) = \prod_{k=t+1}^{s-1} [1 + f(t+1,k)]^{-1}$$

$$= \prod_{k=t+1}^{s-1} [1 + f(t,k)]^{-1} u(t+1,k)$$

$$= \prod_{k=t+1}^{s-1} [1 + f(t-1,k)]^{-1} \prod_{k=t+1}^{s-1} d(t,k)u(t+1,k)$$

$$= \psi^{s-t-1} \prod_{k=t+1}^{s-1} [1 + f(t-1,k)]^{-1} \prod_{k=t+1}^{s-1} u(t,k)u(t+1,k),$$

the same as that in the first case and hence the binomial model is recombining.

CHAPTER 4

CONTINUOUS-TIME STOCHASTIC PROCESSES

4.1 GENERAL DESCRIPTION OF CONTINUOUS-TIME STOCHASTIC PROCESSES

Just as in the discrete-time case, the description of a continuous-time stochastic process begins with a time-dependent information structure or filtration $\{\mathcal{F}_t, 0 \leq t \leq T\}$, on a probability space $(\Omega, \mathcal{F}, P)$. The information structure $\{\mathcal{F}_t\}$ satisfies the usual conditions:

(i) For each t, $\mathcal{F}_t$ is an information structure containing no more information than $\mathcal{F}$, and it represents the information up to time t;

(ii) If $s < t$, $\mathcal{F}_s$ contains no more information than $\mathcal{F}_t$, i.e. all the events in $\mathcal{F}_s$ are events in $\mathcal{F}_t$. In other words, $\{\mathcal{F}_t\}$ is an increasing sequence in t.

A collection of random variables $\{X(t),\ 0 \leq t \leq T\}$ forms a continuous-time stochastic process[1] if for each t, $X(t)$ is a random variable on $\mathcal{F}_t$. The possible values of $X(t)$, $0 \leq t \leq T$, are called the states of the stochastic process. For each t and each outcome $\omega \in \Omega$, $X(t, \omega)$, the value of the random variable $X(t)$ at ω, is a real number. Hence for each fixed ω, $X(t, \omega)$ may be viewed as a real-valued function of t. We call this function a path of the stochastic process.

As in Section 3.1, in practice, we often begin with a collection of random variables $\{X(t), 0 \leq t \leq T\}$ on a probability space $(\Omega, \mathcal{F}, P)$ without having an information structure. In this case, we may use the information structure generated from the random variables $X(t)$, $0 \leq t \leq T$. The information structure at time t, denoted by $\mathcal{B}_t$, is the collection of events observable through $X(s)$, for all $0 \leq s \leq t$, similar to the discrete-time case that is discussed in Section 3.1. The information structure $\{\mathcal{B}_t, 0 \leq t \leq T\}$, obtained this way is referred to as the Borel or natural information structure, and obviously $\{X(t)\}$ is a stochastic process on $\{\mathcal{B}_t\}$.

Most stochastic processes in practical applications have some desirable properties. For example, all the paths of these stochastic processes either are continuous or have only a finite number of jumps over any finite time horizon. Two classes of stochastic processes, Brownian motion and the Poisson process, play a very important role in applications. The former is the building block of stochastic (Ito) calculus and has extensive applications in finance. The latter has many applications in insurance modelling and is familiar to actuaries. Recently, the Poisson process has been used to model stock prices with random dividend payments.

In the next section we introduce Brownian motion and discuss its basic properties.

4.2   BROWNIAN MOTION

In Chapter 3, we discuss extensively a class of discrete-time stochastic processes known as random walks and their properties. As seen in the chapter, random walks have many desirable properties and may be used in financial modelling. In this section, we introduce a class of continuous-time stochastic processes known as Brownian motion processes. We will describe Brownian motion as the limit of a sequence of random walks and hence Brownian motion may be viewed as a continuous-time version of a random walk. Many important properties of Brownian motion as a result are derived naturally from the properties of random walks.

To understand how the transition from a random walk to Brownian motion takes place, we consider a case where an asset price follows a geometric

---

[1] For a well-defined continuous-time stochastic process, some technical conditions need to be imposed. However, for application purposes these conditions are not important.

random walk over a period of time as described in Section 3.3. In this case, the asset price $S(t)$ at time t may be expressed as

$$S(t) = S(0)e^{X_\tau(t)}, \ 0 \leq t \leq T,$$

where $\{X_\tau(t)\}$ is a random walk with step size τ, drift μ, and volatility σ, where τ represents the length of a trading period. See Section 3.3 for details.

Imagine now that trading becomes more and more frequent and eventually continuous trading is achieved. In mathematical terms, the step size of the random walk τ tends to 0. If the limit of $X_\tau(t)$ as $\tau \to 0$, denoted as $W_{\mu,\sigma}(t)$, exists for each t, one obtains a continuous-time stochastic process $\{W_{\mu,\sigma}(t)\}$. In this case, the asset price may be expressed as $S(t) = S(0)e^{W_{\mu,\sigma}(t)}$, $0 \leq t \leq T$. In what follows, we will show that such a limiting stochastic process does exist, and will identify this stochastic process. Our approach involves (i) calculating the Laplace transform of the random walk $X_\tau(t)$ for each t, (ii) finding the limit of the above Laplace transform, and (iii) identifying the limit as the Laplace transform of some random variable, say $W_{\mu,\sigma}(t)$. If this approach is applicable, a result in probability theory tells us that $W_{\mu,\sigma}(t)$ is indeed the limit of $X_\tau(t)$.

We now begin with the Laplace transform of the random walk. Let $\tilde{f}_\tau(z,t)$ be the Laplace transform of $X_\tau(t)$. Recalling that for each fixed t, $X_\tau(t)$ is a binomial random variable, one obtains from Example 2.5

$$\tilde{f}_\tau(z,t) = \left[qe^{-h_1 z} + (1-q)e^{h_2 z} \right]^{\bar{t}},$$

where $\bar{t} = t/\tau$.

Also, recall from Section 3.2 that if $E\{X_\tau(t)\} = \mu t$ and $Var\{X_\tau(t)\} = \sigma^2 t$, then we may choose

$$h_1 = \mu\tau + \sigma\sqrt{\tau}, \ h_2 = -\mu\tau + \sigma\sqrt{\tau}, \ q = \frac{1}{2}.$$

In this case,

$$\tilde{f}_\tau(z,t) = e^{-\mu t z} \left(\frac{e^{-\sigma\sqrt{\tau}z} + e^{\sigma\sqrt{\tau}z}}{2} \right)^{t/\tau}.$$

Next, we compute the limit of the Laplace transform $\tilde{f}_\tau(z,t)$. If we can show

$$\lim_{\tau \to 0} \left(\frac{e^{-\sigma\sqrt{\tau}z} + e^{\sigma\sqrt{\tau}z}}{2} \right)^{1/\tau} = e^{\frac{1}{2}\sigma^2 z^2}, \tag{4.1}$$

then

$$\lim_{\tau \to 0} \tilde{f}_\tau(z,t) = e^{-\mu t z + \frac{1}{2}\sigma^2 t z^2}.$$

By taking the natural logarithm of the left-hand side of (4.1), we have

$$\frac{1}{\tau} \ln \left(\frac{e^{-\sigma\sqrt{\tau}z} + e^{\sigma\sqrt{\tau}z}}{2} \right). \tag{4.2}$$

In order to apply L'Hospital's rule to (4.2), differentiate $\ln \left(\frac{e^{-\sigma\sqrt{\tau}z}+e^{\sigma\sqrt{\tau}z}}{2} \right)$ with respect to τ, which yields

$$\frac{\sigma z}{2\sqrt{\tau}} \left(\frac{-e^{-\sigma\sqrt{\tau}z} + e^{\sigma\sqrt{\tau}z}}{e^{-\sigma\sqrt{\tau}z} + e^{\sigma\sqrt{\tau}z}} \right).$$

Expanding $e^{-\sigma\sqrt{\tau}z}$ and $e^{\sigma\sqrt{\tau}z}$ in terms of $\sqrt{\tau}$ at $\tau = 0$ in a Taylor series yields

$$\lim_{\tau\to 0} \frac{-e^{-\sigma\sqrt{\tau}z} + e^{\sigma\sqrt{\tau}z}}{\sqrt{\tau}} = \lim_{\tau\to 0} \frac{2\sigma\sqrt{\tau}z + o(\tau)}{\sqrt{\tau}} = 2\sigma z,$$

where $o(\tau)$ represents the higher-order terms with respect to τ. Thus, it follows from L'Hospital's rule that

$$\lim_{\tau\to 0} \frac{1}{\tau} \ln \left(\frac{e^{-\sigma\sqrt{\tau}z} + e^{\sigma\sqrt{\tau}z}}{2} \right) = \lim_{\tau\to 0} \frac{\sigma z}{2\sqrt{\tau}} \left(\frac{-e^{-\sigma\sqrt{\tau}z} + e^{\sigma\sqrt{\tau}z}}{e^{-\sigma\sqrt{\tau}z} + e^{\sigma\sqrt{\tau}z}} \right) = \frac{1}{2}\sigma^2 z^2.$$

Thus

$$\lim_{\tau\to 0} \left(\frac{e^{-\sigma\sqrt{\tau}z} + e^{\sigma\sqrt{\tau}z}}{2} \right)^{1/\tau} = e^{\frac{1}{2}\sigma^2 z^2}.$$

Therefore,

$$\lim_{\tau\to 0} \hat{f}_\tau(z,t) = e^{-\mu tz + \frac{1}{2}\sigma^2 t z^2}.$$

Since the limit is a continuous function, by Levy's Convergence Theorem (Williams (1994), Chapter 18) $X_\tau(t)$ converges weakly[2] to a random variable $W_{\mu,\sigma}(t)$. Comparing it with the Laplace transform of a normal random variable in Example 2.7, we see that $W_{\mu,\sigma}(t)$ is a normal random variable with mean μt and variance $\sigma^2 t$.

Obviously, the limiting stochastic process $\{W_{\mu,\sigma}(t)\}$ will inherit some properties that the random walk $\{X_\tau(t)\}$ possesses. For example, $W_{\mu,\sigma}(0) =$

---

[2] Weak convergence is defined as follows: a sequence of random variable X_n converges weakly to a random variable X if for any bounded continuous function $h(x)$, $\lim_{n\to\infty} E\{h(X_n)\} = E\{h(X)\}$. It can be shown that weak convergence is equivalent to one of the following:

   (i) The distribution function of X_n converges to the distribution function of X at any continuous point;

   (ii) The Laplace transform of X_n converges to the Laplace transform of X as long as the Laplace transform of X is continuous at $z = 0$.

0, $E\{W_{\mu,\sigma}(t)\} = \mu t$, and $Var\{W_{\mu,\sigma}(t)\} = \sigma^2 t$. Since $X_\tau(t)$ has independent increments, so does $W_{\mu,\sigma}(t)$. Thus, for any partition $0 < t_1 < t_2 < \cdots < t_j < t$,

$$W_{\mu,\sigma}(t_1), W_{\mu,\sigma}(t_2) - W_{\mu,\sigma}(t_1), \cdots, W_{\mu,\sigma}(t) - W_{\mu,\sigma}(t_j)$$

are independent. Thus, the limiting stochastic process $\{W_{\mu,\sigma}(t)\}$ has the following properties.

1. It has independent increments. In other words, for any partition $0 < t_1 < t_2 < \cdots < t_j < t$,

$$W_{\mu,\sigma}(t_1), W_{\mu,\sigma}(t_2) - W_{\mu,\sigma}(t_1), \cdots, W_{\mu,\sigma}(t) - W_{\mu,\sigma}(t_j)$$

are independent. An implication of this property is that $\{W_{\mu,\sigma}(t)\}$ has the Markov property, i.e. the future value of this process depends only on the current value but not on the values in the past. This property can extend to a function of the Brownian motion, and one has the following useful corollary: for any function $h(w)$,

$$E\{h(W_{\mu,\sigma}(s)) \mid W_{\mu,\sigma}(u), 0 \le u \le t\} = E\{h(W_{\mu,\sigma}(s)) \mid W_{\mu,\sigma}(t)\}.$$

2. For any ordered pair $s > t$, the random variable $W_{\mu,\sigma}(s) - W_{\mu,\sigma}(t)$ is normal with

$$
\begin{align}
E\{W_{\mu,\sigma}(s) - W_{\mu,\sigma}(t)\} &= \mu(s - t), \tag{4.3} \\
Var\{W_{\mu,\sigma}(s) - W_{\mu,\sigma}(t)\} &= \sigma^2(s - t). \tag{4.4}
\end{align}
$$

The normality comes from the fact that the difference of two normal random variables is still normal. The equation (4.3) is obvious. From the independent increments property

$$
\begin{align}
&Cov(W_{\mu,\sigma}(s), W_{\mu,\sigma}(t)) \\
=~&Cov(W_{\mu,\sigma}(s) - W_{\mu,\sigma}(t), W_{\mu,\sigma}(t)) + Var\{W_{\mu,\sigma}(t)\} \\
=~&Var\{W_{\mu,\sigma}(t)\} = \sigma^2 t.
\end{align}
$$

Thus

$$
\begin{align}
&Var\{W_{\mu,\sigma}(s) - W_{\mu,\sigma}(t)\} \\
=~&Var\{W_{\mu,\sigma}(s)\} - 2Cov(W_{\mu,\sigma}(s), W_{\mu,\sigma}(t)) + Var\{W_{\mu,\sigma}(t)\} \\
=~&Var\{W_{\mu,\sigma}(s)\} - Var\{W_{\mu,\sigma}(t)\} \\
=~&\sigma^2(s - t).
\end{align}
$$

We now formally introduce Brownian motion as follows. A stochastic process $\{W_{\mu,\sigma}(t)\}$ is called Brownian motion (or Wiener process) with drift μ

and infinitesimal variance σ^2 (or volatility σ) if it satisfies the two properties above. Brownian motion $\{W_{0,1}(t)\}$, i.e. $\mu = 0$ and $\sigma = 1$, is called standard Brownian motion. Hereafter, we will always use $\{W_{\mu,\sigma}(t)\}$ for a Brownian motion process with drift μ and volatility σ, and $\{W(t)\}$ for standard Brownian motion.

It can be shown that Brownian motion has the next two properties, although they are difficult to prove.

3. All the paths (with probability one) of Brownian motion are continuous.

4. All the paths of Brownian motion are nowhere differentiable[3].

Property 3 says that if the position of an object follows Brownian motion $\{W_{\mu,\sigma}(t)\}$, then this object moves continuously over time with no jumps. However, property 4 indicates that this object takes sharp turns at every moment.

More properties can be obtained now. It is easy to see that any linear combination of Brownian motion processes is Brownian motion. For example, $\{-W_{\mu,\sigma}(t)\}$ is Brownian motion with drift $-\mu$ and the same volatility σ. For Brownian motion $\{W_{\mu,\sigma}(t)\}$, since $E\left\{\frac{1}{\sigma}(W_{\mu,\sigma}(t) - \mu t)\right\} = 0$ and $Var\left\{\frac{1}{\sigma}(W_{\mu,\sigma}(t) - \mu t)\right\} = t$, the process $\frac{1}{\sigma}(W_{\mu,\sigma}(t) - \mu t)$ is standard Brownian motion, i.e., $W(t) = \frac{1}{\sigma}(W_{\mu,\sigma}(t) - \mu t)$. For this reason, any Brownian motion process starting at 0 (i.e. $W(0) = 0$) is written in the form of $\{\mu t + \sigma W(t)\}$. Moreover, Brownian motion is preserved under shift transformation, scale transformation and time inversion transformation. More precisely, if $\{W(t)\}$ is standard Brownian motion, then for $h > 0$, the shifted process $W_1(t) = W(t + h) - W(h)$, the scaled process $W_2(t) = hW(t/h^2)$, and the time-inverted process $W_3(t) = tW(1/t)$, with $W_3(0) = 0$, all are standard Brownian motion processes. The proof of these facts is left to interested readers. Finally, define $W_4(t) = |W(t)|$. The process $\{W_4(t)\}$ is called Brownian motion reflected at the origin. It can be shown that $E\{W_4(t)\} = \sqrt{2t/\pi}$ and $Var\{W_4(t)\} = (1 - 2/\pi)t$. Again, we leave the proof to interested readers.

Sometimes we need to deal with a Brownian motion process starting at a point, say x, away from zero. Standard Brownian motion in this case is $\{x + W(t)\}$. We call it standard Brownian motion starting at x and denote it as $\{W(t),\ W(0) = x\}$.

Some sample paths of three Brownian motion processes with different drifts and volatilities are given in Figures 4.1 to 4.5 at the top of the next five pages. These figures show the impact of the drift and volatility to the Brownian motion processes.

---

[3]It is difficult to imagine a continuous but nowhere differentiable function. To better understand this type of functions, readers may consult Gelbaum and Olmstead (1964) in which a continuous but nowhere differentiable function is constructed.

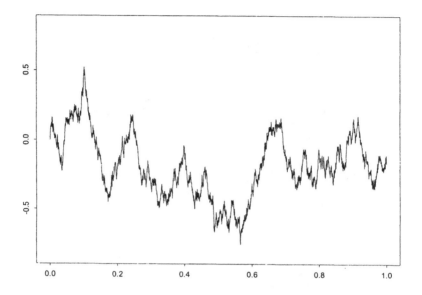

Figure 4.1. A sample path of standard Brownian motion ($\mu = 0$ and $\sigma = 1$).

We now return to the security price process $\{S(t)\}$ discussed at the beginning of this section. Since the random walks approach Brownian motion with drift μ and volatility σ when $\tau \rightarrow 0$, the continuous-time price process is of the form

$$S(t) = S(0)e^{\mu t + \sigma W(t)}, \qquad (4.5)$$

where $\{W(t)\}$ is standard Brownian motion. A stochastic process of this form is referred to as *geometric Brownian motion*. It is easy to see that

$$E\{S(t)\} = S(0)e^{(\mu + \frac{1}{2}\sigma^2)t}. \qquad (4.6)$$

We should point out that $\mu + \frac{1}{2}\sigma^2$ represents the expected rate of return compounded continuously and for this reason it is often called the drift of geometric Brownian motion. The parameter σ is still called the volatility of geometric Brownian motion.

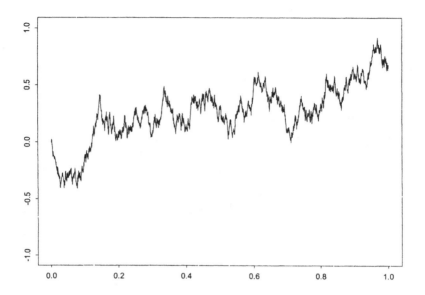

Figure 4.2.   A sample path of Brownian motion with $\mu = 1$ and $\sigma = 1$.

4.3   THE REFLECTION PRINCIPLE AND BARRIER HITTING PROBABILITIES

In this section, we discuss a very unique and important property of standard Brownian motion: the Reflection Principle. Roughly speaking, the Reflection Principle says that a path obtained by reflecting a portion of a standard Brownian motion path (with any initial value, 0 or away from 0) about a horizontal line is either another path of the same standard Brownian motion process or a path of another standard Brownian motion process. The Reflection Principle is a very useful tool for it is used to derive many hitting time distributions of Brownian motion.

Theorem 4.1 (The Reflection Principle)   Let $\{W(t)\}$ be standard Brownian motion and let a and h be two nonnegative real numbers. Then,

$$P\left\{\max_{0<t\leq s} W(t) \geq a, W(s) \geq a+h\right\}$$

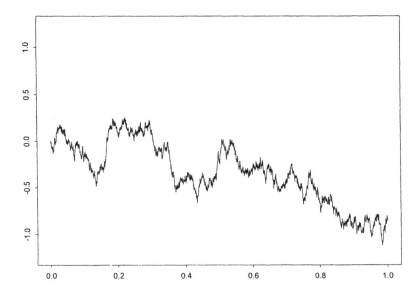

Figure 4.3. A sample path of Brownian motion with $\mu = -1$ and $\sigma = 1$.

$$= P\left\{\max_{0<t\leq s} W(t) \geq a, W(s) \leq a - h\right\}, \qquad (4.7)$$

where we use the convention that the set $\{A, B\}$ means A and B. The proof of the Reflection Principle is omitted. We refer interested readers to Durrett (1996), p. 25. The following is another version of the Reflection Principle.

Corollary 4.1 Let $[a + h_1, a + h_2]$, $h_2 \geq h_1 \geq 0$, be any interval above the horizontal line $x = a$, with symmetric interval $[a - h_2, a - h_1]$ below the line $x = a$, we have

$$P\left\{\max_{0<t\leq s} W(t) \geq a, W(s) \in [a + h_1, a + h_2]\right\}$$
$$= P\left\{\max_{0<l\leq s} W(t) \geq a, W(s) \in [a - h_2, a - h_1]\right\}. \qquad (4.8)$$

Proof: Obviously,

$$P\left\{\max_{0<t\leq s} W(t) \geq a, W(s) \in [a + h_1, a + h_2]\right\}$$

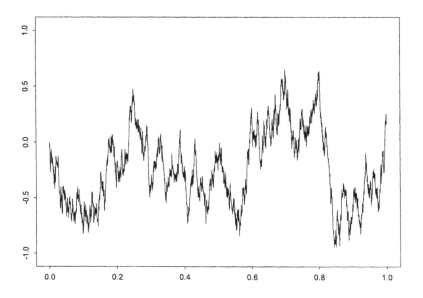

Figure 4.4. A sample path of Brownian motion with $\mu = 0$ and $\sigma = 2$.

$$= \; P\Big\{ \max_{0<t\leq s} W(t) \geq a, W(s) \geq a + h_1 \Big\}$$

$$- \; P\Big\{ \max_{0<t\leq s} W(t) \geq a, W(s) \geq a + h_2 \Big\}.$$

By Theorem 4.1,

$$P\Big\{ \max_{0<t\leq s} W(t) \geq a, W(s) \geq a+h_1 \Big\} = P\Big\{ \max_{0<t\leq s} W(t) \geq a, W(s) \leq a-h_1 \Big\},$$

and

$$P\Big\{ \max_{0<t\leq s} W(t) \geq a, W(s) \geq a+h_2 \Big\} = P\Big\{ \max_{0<t\leq s} W(t) \geq a, W(s) \leq a-h_2 \Big\}.$$

Thus

$$P\Big\{ \max_{0<t\leq s} W(t) \geq a, W(s) \in [a + h_1, \, a + h_2] \Big\}$$

$$= \; P\Big\{ \max_{0<t\leq s} W(t) \geq a, W(s) \leq a - h_1 \Big\}$$

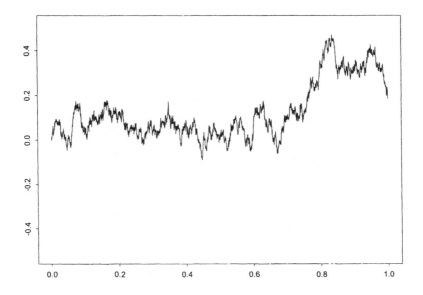

Figure 4.5. A sample path of Brownian motion with $\mu = 0$ and $\sigma = 0.5$.

$$- \quad P\left\{ \max_{0 < t \le s} W(t) \ge a, W(s) \le a - h_2 \right\}$$
$$= \quad P\left\{ \max_{0 < t \le s} W(t) \ge a, W(s) \in [a - h_2, \ a - h_1] \right\}.$$

$\square$

Interpretation of the Reflection Principle

Note that $\left\{ \max_{0<t\le s} W(t) \ge a \right\}$ represents the event that the path of standard Brownian motion $\{W(t)\}$ hits the horizontal barrier $x = a$ before or at time s. Hence, $\left\{ \max_{0<t\le s} W(t) \ge a, W(s) \in [a+h_1, \ a+h_2] \right\}$ is the event that the path of standard Brownian motion $\{W(t)\}$ hits the horizontal barrier $x = a$ before or at time s and then reaches the interval $[a + h_1, \ a + h_2]$ at time s. Similarly, $\left\{ \max_{0<t\le s} W(t) \ge a, W(s) \in [a - h_2, \ a - h_1] \right\}$ is the event that the path of standard Brownian motion $\{W(t)\}$ hits the barrier $x = a$ and then reaches the interval $[a - h_2, \ a - h_1]$. The Reflection Principle (i.e. Corollary 4.1) thus says that the probability that standard Brownian motion

hits a barrier and reaches an interval is the same as the probability that it hits the barrier and reaches the mirror image of the interval about the barrier. Since h_1, h_2 are arbitrary, we may let the interval $[a + h_1,\ a + h_2]$ shrink to a single point, say y, i.e $h_1 = h_2 = y - a$. Then the Reflection Principle says that if a path hits the barrier $x = a$ at some time $t_a < s$ and reaches the value y at time s, there is another path that is identical to the first path up to the hitting time t_a and is then the mirror image of the first path about the barrier $x = a$ after t_a. In other words, if we reflect the portion of a path of standard Brownian motion after a hitting point, we obtain another path of standard Brownian motion. Figure 4.6 on p. 109 shows a path of standard Brownian motion being reflected about the barrier $a = 0.135$ after the hitting time $t_a = 0.72$. Another way to obtain a path of standard Brownian motion is to reflect the portion of a path of another Brownian motion process which starts at time 0 and ends at some hitting point. To see this, note that $\{2a - W(t)\}$ is another standard Brownian motion process and it is the mirror image of $\{W(t)\}$ about the barrier $x = a$. If a path of $\{W(t)\}$ hits the barrier $x = a$, say at some time $t_a < s$, so does the corresponding path of $\{2a - W(t)\}$. Thus, we may apply the Reflection Principle to the path of $\{2a - W(t)\}$ by reflecting its portion after time t_a. The resulting path is a path of $\{2a - W(t)\}$ as we just described. Note that the portion of this path after reflection is exactly the same as the corresponding portion of the path of $\{W(t)\}$ after time t_a. Hence, the path obtained by reflecting the path of $\{2a - W(t)\}$ after time t_a is such that it is a mirror image of the path of $\{W(t)\}$ up to time t_a and the path of $\{W(t)\}$ itself after time t_a. In other words, if we reflect a path of $\{W(t)\}$ before a hitting time, we obtain a path of $\{2a -\!- W(t)\}$. Figure 4.7 on p. 110 shows how a path of a standard Brownian motion process is generated from reflecting a path of another standard Brownian motion process up to a hitting point.

As we mentioned at the beginning of this section, a very important application of the Reflection Principle is to calculate various barrier hitting time distributions of Brownian motion. Barrier hitting time distributions are of central importance because they are used in the pricing of barrier options, digital options, and other exotic options in finance.

We now consider two fairly simple cases to illustrate how to use the Reflection Principle to calculate barrier hitting time distributions.

First Passage Time

Consider the following random variable

$$T_a = \inf\{t > 0;\ W(t) = a\}. \tag{4.9}$$

The random variable T_a is the first time standard Brownian motion $\{W(t)\}$ hits the barrier $x = a$ and is called a first passage time of $\{W(t)\}$.

Original path of W(t)

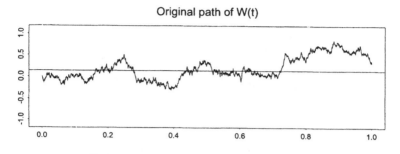

Path reflected at barrier x=0.135 at time t=0.72

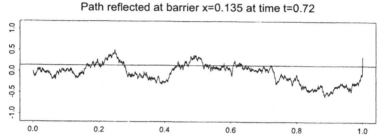

Figure 4.6. A path of standard Brownian motion reflected after hitting.

First, assume $a > 0$, i.e the barrier $x = a$ is an upper barrier. To find the distribution of T_a, we examine the event that the hitting time T_a is less than or equal to time t, i.e. $\{T_a \leq t\}$. This event is the same as the event that the maximum of $\{W(s)\}$ between times 0 and t is at least a. That is,

$$\{T_a \leq t\} = \left\{ \max_{0 < s \leq t} W(s) \geq a \right\}.$$

Thus, the distribution function of T_a, denoted as $F_a(t) = P\{T_a \leq t\}$ for $a > 0$ can be derived as follows.

$$
\begin{aligned}
P\{T_a \leq t\} &= P\left\{ \max_{0 < s \leq t} W(s) \geq a \right\} \\
&= P\left\{ \max_{0 < s \leq t} W(s) \geq a, W(t) \geq a \right\} \\
&+ P\left\{ \max_{0 < s \leq t} W(s) \geq a, W(t) \leq a \right\} \\
&\quad (\text{since } P\{W(t) = a\} = 0) \\
&= 2P\left\{ \max_{0 < s \leq t} W(s) \geq a, W(t) \geq a \right\}
\end{aligned}
$$

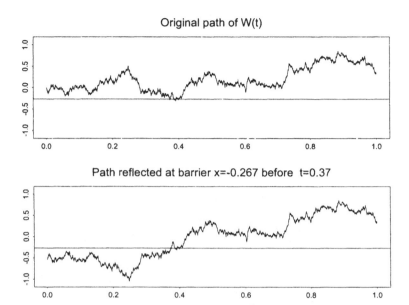

Figure 4.7. A path of standard Brownian motion reflected before hitting.

$$\text{(by the Reflection Principle)}$$
$$= \ 2P\{W(t) \ge a\}.$$
$$\left(\text{since } \{W(t) \ge a\} \subseteq \left\{\max_{0 < s \le t} W(s) \ge a\right\}\right)$$

Since $W(t)$ is normally distributed with mean 0 and variance t, one has

$$P\{T_a \le t\} = 2P\{W(t) \ge a\} = \sqrt{\frac{2}{\pi t}} \int_a^\infty e^{-\frac{1}{2t}x^2}\, dx = \sqrt{\frac{2}{\pi}} \int_{a/\sqrt{t}}^\infty e^{-\frac{1}{2}x^2}\, dx.$$

If the barrier $x = a$ is a lower barrier, i.e. $a < 0$, then

$$P\{T_a \le t\} = \sqrt{\frac{2}{\pi}} \int_{-a/\sqrt{t}}^\infty e^{-\frac{1}{2}x^2}\, dx.$$

This can be obtained by considering $-W(t)$ instead of $W(t)$ with the barrier $x = -a$. Since $-W(t)$ is also a standard Brownian motion process, the formula above applies.

It is easy to see from the above that T_a is finite (e.g. $P\{T_a = \infty\} = 0$) since

$$P\{T_a < \infty\} = \lim_{t \to \infty} P\{T_a \le t\} = \sqrt{\frac{2}{\pi}} \int_0^\infty e^{-\frac{1}{2}x^2} dx = 1.$$

The intuitive meaning is that for any horizontal barrier, every path of standard Brownian motion will hit the barrier sooner or later.

Now, let $f_a(t)$ be the density function of T_a. We have

$$f_a(t) = \frac{dF_a(t)}{dt} = \frac{|a|}{\sqrt{2\pi t^3}} e^{-\frac{a^2}{2t}}, \quad t > 0. \tag{4.10}$$

This distribution is called the one-sided stable distribution of index $\frac{1}{2}$, which can be obtained as a limit of inverse Gaussian distributions. To see this, recall (Example 2.8) that the density of an inverse Gaussian distribution is given by

$$f_{IG}(t) = \frac{\alpha}{\sqrt{2\pi\beta t^3}} e^{-\frac{1}{2\beta t}(\beta t - \alpha)^2}, \quad t > 0.$$

Letting $\alpha = |a|\sqrt{\beta}$ and $\beta \to 0$, we obtain (4.10).

Single Barrier

We now consider the distribution of standard Brownian motion $W(T)$ at a fixed time T given that its path hits a barrier earlier. Without taking the barrier hitting situation into account, the distribution of $W(T)$ is normal with mean 0 and variance T. However, the distribution of $W(T)$ is quite different given that its path hits a barrier earlier, as we will show in the following. The tool used to calculate such a distribution is again the Reflection Principle.

Let $g_a(x)$, $a > 0$ be the density function of $W(T)$, given that $W(t) = a$, for some $0 < t \le T$. Thus $g_a(y)dy$ is the probability that a path hits the barrier $x = a$ and then reaches point y at time T. It is easy to see that $g_a(y)$ is a defective density function (a density function is called defective if its integral is less than one) since not all paths will hit the barrier by time T.

Obviously, for $y \ge a$, any path to reach y at T will hit the barrier $x = a$ before or at T. Hence, the probability that a path hits $x = a$ and reaches y at T is the same as the probability that a path reaches y at T. The normality of $W(T)$ implies that this probability is

$$g_a(y)dy = \frac{1}{\sqrt{2\pi T}} e^{-\frac{y^2}{2T}} dy, \quad y \ge a.$$

For the case $y < a$, we use the Reflection Principle. The probability of a path that hits the barrier a and then reaches y at T is equal to the probability of a path that starts at $2a$, hits the barrier a, and then reaches y at T. This is the case when we reflect the portion of the path between time 0 and its hitting time. But the probability of a path that starts at $2a$, hits the barrier a, and then reaches y at T is the same as the probability that a path starts at $2a$ and then reaches y at T, since the barrier is between the starting point $2a$ and the value y. Similar to the case $y \geq a$, this probability is

$$g_a(y)dy = \frac{1}{\sqrt{2\pi T}} e^{-\frac{(y-2a)^2}{2T}} dy, \ y < a.$$

This is because the distribution of $2a + W(T)$ is normal with mean $2a$ and variance T. Together, we have shown that the density $g_a(y)$ is given by

$$g_a(y) = \begin{cases} \frac{1}{\sqrt{2\pi T}} e^{-\frac{(y-2a)^2}{2T}}, & y < a \\ \frac{1}{\sqrt{2\pi T}} e^{-\frac{y^2}{2T}}, & y \geq a. \end{cases} \tag{4.11}$$

4.4 THE POISSON PROCESS AND COMPOUND POISSON PROCESS

In this section, we introduce Poisson and compound Poisson processes. Poisson and compound Poisson processes are used extensively in insurance modelling. These processes are well understood by actuaries. In the previous section we introduced Brownian motion as the limit of a sequence of random walks. This approach is common in finance but may be difficult for actuaries. An alternative approach is to treat Brownian motion as the limit of a sequence of Poisson processes. An advantage to this approach is that it gives us a better understanding of the properties of Brownian motion. Moreover, one may use a Poisson process to approximate a Brownian motion when necessary. We will discuss this approach in detail in this section. Other than using Poisson processes to describe and approximate Brownian motions, Poisson and compound Poisson processes are useful in their own right in financial modelling. They are often used to model a dividend payment process when the time and the amount of the dividend payments are random.

Poisson Process

A continuous-time counting stochastic process[4] $\{N(t), \ t \geq 0\}$ is called a Poisson process if it satisfies the following conditions:

1. $N(0) = 0$;

2. $\{N(t)\}$ has independent increments, i.e. for any partition $0 < t_1 < t_2 < \cdots < t_j < t$, the increments

$$N(t_1), N(t_2) - N(t_1), \cdots, N(t) - N(t_j)$$

---

[4]A counting stochastic process is such that its values are non-negative integers and and each path is nondecreasing.

are independent;

3. There is a $\lambda > 0$ such that for each ordered pair $s < t$, the increment $N(t) - N(s)$ is a Poisson random variable with parameter $\lambda(t - s)$. That is, for $n = 0, 1, 2, \cdots$,

$$P\{N(t) - N(s) = n\} = e^{-\lambda(t-s)} \frac{[\lambda(t - s)]^n}{n!}. \qquad (4.12)$$

The parameter λ is called the intensity of the Poisson process and represents the average number of occurrences per unit time. An immediate implication of the above is that for each t, $N(t)$ has a Poisson distribution with parameter λt. Thus, from Example 2.6

$$E\{N(t)\} = Var\{N(t)\} = \lambda t. \qquad (4.13)$$

If we interpret $\{N(t)\}$ as the number of claims occurring by time t, the parameter λ is the average number of claims per unit time. The independent increments property implies that the time of the next claim is independent of the time of the claims that occurred up to the current time. It can also be shown that the inter-occurrence time between two consecutive claims is exponentially distributed with parameter λ.

Compound Poisson Process

A compound Poisson process $\{S(t),\ t \geq 0\}$ is a random sum process given by

$$S(t) = Y_1 + Y_2 + \cdots + Y_{N(t)}, \qquad (4.14)$$

with $S(t) = 0$, if $N(t) = 0$. Here $\{N(t)\}$ is a Poisson process with intensity λ. The sequence of random variables $Y_1, Y_2, \cdots$ are independent and identically distributed, and independent of $\{N(t)\}$. We may interpret $Y_1, Y_2, \cdots$ as consecutive individual claim amounts. Thus, $S(t)$ represents the aggregate claims amount at time t. Figure 4.8. shows a path of this process. A simple calculation yields

$$E\{S(t)\} = \lambda t E\{Y_1\}, \quad Var\{S(t)\} = \lambda t E\{Y_1^2\}. \qquad (4.15)$$

Also see Ross (1993, pp.239-241). It is easy to see that the Laplace transform of $S(t)$ is

$$\tilde{f}_S(z) = e^{\lambda t [\tilde{f}_Y(z) - 1]}, \qquad (4.16)$$

where $\tilde{f}_Y(z)$ is the Laplace transform of Y.

An important property of the Poisson and compound Poisson processes is that they are preserved under addition. In other words, the sum of independent Poisson processes is a Poisson process and the sum of independent

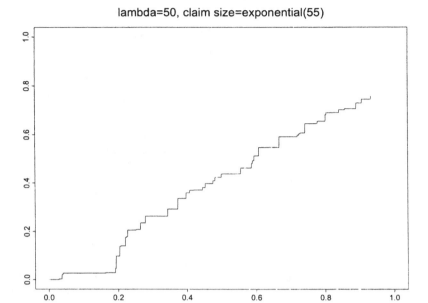

Figure 4.8.   A sample path of a compound Poisson process.

compound Poisson processes is a compound Poisson process. More precisely, if $\{N_1(t)\}, \{N_2(t)\}, \cdots, \{N_m(t)\}$ are m independent Poisson processes with parameters $\lambda_1, \lambda_2, \cdots, \lambda_m$, respectively, then the sum

$$N(t) = N_1(t) + N_2(t) + \cdots + N_m(t) \tag{4.17}$$

is a Poisson process with parameter

$$\lambda = \lambda_1 + \lambda_2 + \cdots + \lambda_m. \tag{4.18}$$

Furthermore, if $\{S_1(t)\}, \{S_2(t)\}, \cdots, \{S_m(t)\}$ are m independent compound Poisson processes with parameters $\lambda_1, \lambda_2, \cdots, \lambda_m$ and with the claim amount distribution functions $F_1(y), F_2(y), \cdots, F_m(y)$, respectively, then the sum

$$S(t) = S_1(t) + S_2(t) + \cdots + S_m(t) \tag{4.19}$$

is a compound Poisson process with Poisson parameter λ given in (4.18) and the claim amount distribution function

$$F(y) = \frac{\lambda_1}{\lambda} F_1(y) + \frac{\lambda_2}{\lambda} F_2(y) + \cdots + \frac{\lambda_m}{\lambda} F_m(y). \tag{4.20}$$

To verify that $\{N(t)\}$ in (4.17) is a Poisson process and $\{S(t)\}$ in (4.19) is a compound Poisson process, we simply examine the Laplace transform of $\{S(t)\}$, as $\{N(t)\}$ may be viewed as a special case of $\{S(t)\}$ where the claim amount random variables $Y = 1$. It follows from (4.16) that

$$
\begin{aligned}
\tilde{f}_S(z) &= E\left\{e^{-zS_1(t) - zS_2(t) - \cdots - zS_m(t)}\right\} \\
&= E\left\{e^{-zS_1(t)}\right\} E\left\{e^{-zS_2(t)}\right\} \cdots E\left\{e^{-zS_m(t)}\right\} \\
&= e^{\lambda_1 t(\tilde{f}_1(z) - 1)} e^{\lambda_2 t(\tilde{f}_2(z) - 1)} \cdots e^{\lambda_m t(\tilde{f}_m(z) - 1)} \\
&= e^{\lambda t \left[\frac{\lambda_1}{\lambda}\tilde{f}_1(z) + \frac{\lambda_2}{\lambda}\tilde{f}_2(z) + \cdots + \frac{\lambda_m}{\lambda}\tilde{f}_m(z) - 1\right]},
\end{aligned}
$$

where λ given in (4.18) and $\tilde{f}_i(z), i = 1, 2, \cdots, m$, are the Laplace transforms of $F_i(y), i = 1, 2, \cdots, m$. It is easy to see that

$$
\tilde{f}_Y(z) = \frac{\lambda_1}{\lambda}\tilde{f}_1(z) + \frac{\lambda_2}{\lambda}\tilde{f}_2(z) + \cdots + \frac{\lambda_m}{\lambda}\tilde{f}_m(z)
$$

is the Laplace transform of $F(y)$ given in (4.20). Comparing with (4.16), we can conclude that (4.19) is a compound Poisson process.

We now consider approximation of a Brownian motion with Poisson processes. The following approach was suggested by Gerber and Shiu in a series of papers (for example, see Gerber and Shiu (1996)).

Consider a stochastic process

$$
X_\tau(t) = \tau N(t) - ct, \ t \geq 0, \tag{4.21}
$$

where $\{N(t)\}$ is a Poisson process with parameter λ (Figure 4.9.). This process is called a shifted Poisson process and can be viewed as a total loss to an insurance company up to time t where c represents the rate of premuim income and τ the amount of each claim which is assumed to be a constant. Obviously, the stochastic process $\{X_\tau(t)\}$ has independent increments.

In order to approximate Brownian motion with drift μ and volatility σ, we need to first choose proper parameters for the shifted Poisson process $\{X_\tau(t)\}$, and then verify that under that choice of parameters the process $\{X_\tau(t)\}$ approaches Brownian motion when $\tau \to 0$.

In order to choose parameters properly, we calculate the mean and variance of $X_\tau(t)$ for a fixed t. A simple calculation yields

$$
E\{X_\tau(t)\} = (\tau\lambda - c)t, \ Var\{X_\tau(t)\} = \tau^2 \lambda t.
$$

By matching the means and variances of $W(t)$ and $X_\tau(t)$ we choose parameters λ and c to satisfy

$$
\tau\lambda - c = \mu, \ \tau^2 \lambda = \sigma^2.
$$

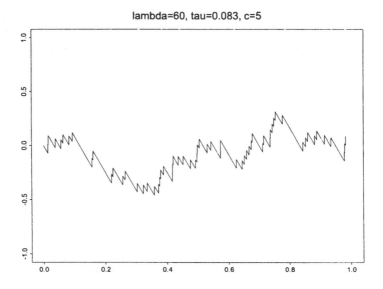

lambda=60, tau=0.083, c=5

Figure 4.9. A sample path of the shifted Poisson process $\{X_\tau(t)\}$.

Solving the above equations yields

$$\lambda = \left(\frac{\sigma}{\tau}\right)^2, \quad c = \frac{\sigma^2}{\tau} - \mu. \tag{4.22}$$

To verify that $\{X_\tau(t)\}$ approaches Brownian motion with drift μ and volatility σ as $\tau \to \infty$, we examine the Laplace transform $\tilde{f}_\tau(z)$ of $X_\tau(t)$. It follows from (4.22) and (2.13) that

$$
\begin{aligned}
\tilde{f}_\tau(z) &= E\left\{e^{-zX_\tau(t)}\right\} = E\left\{e^{-z(\tau N(t) - ct)}\right\} \\
&= e^{ctz} E\left\{e^{-z\tau N(t)}\right\} = e^{ctz} e^{\lambda t(e^{-z\tau} - 1)} \\
&= e^{-\mu z t + \left[\sigma^2 z/\tau + \sigma^2(e^{-z\tau} - 1)/\tau^2\right] t}
\end{aligned}
$$

Let $\tau \to 0$. Since $e^{-z\tau} - 1 = -z\tau + \frac{1}{2}z^2\tau^2 + o(\tau^2)$,

$$
\begin{aligned}
\lim_{\tau \to 0} \tilde{f}_\tau(z) &= \lim_{\tau \to 0} e^{-\mu z t + \left[\sigma^2 z/\tau + \sigma^2(e^{-z\tau} - 1)/\tau^2\right] t} \\
&= e^{-\mu z t + \frac{1}{2}\sigma^2 z^2 t}
\end{aligned}
$$

Hence, the limit is indeed a Brownian motion with drift μ and volatility σ.

4.5 MARTINGALES

The notion of martingale for continuous-time stochastic processes is very similar to the discrete-time case. Suppose that $\{M(t)\}$ is a continuous-time stochastic process on a probability space $(\Omega, \mathcal{F}, \mathcal{F}_t, P)$, where $\{\mathcal{F}_t, t \geq 0, \}$ is an information structure. $\{M(t)\}$ is a martingale if $E\{|M(t)|\} < \infty$, $t \geq 0$, and for any $s > t$

$$E\{M(s) \mid \mathcal{F}_t\} = M(t). \tag{4.23}$$

Equivalent definitions are

$$E\{M(s) - M(t) \mid \mathcal{F}_t\} = 0, \tag{4.24}$$

and

$$E\{M(s)/M(t) \mid \mathcal{F}_t\} = 1, \text{ if } M(t) > 0. \tag{4.25}$$

As in the discrete case, we have
(i) For any t, $E\{M(t)\} = M(0)$, i.e., the expectation of a martingale remains constant over time.
(ii) For any random variable Y on $\mathcal{F}_T$ with $E\{|Y|\} < \infty$, the stochastic process $\{M(t) = E\{Y \mid \mathcal{F}_t\}\}$, $0 \leq t \leq T$, is a martingale.
(iii) If a martingale $\{M(t)\}$ has independent increments (and hence is a Markov process), then $E\{M(s) \mid M(t)\} = M(t)$, i.e., the information structure $\mathcal{F}_t$ can be replaced by the information structure generated by $M(t)$.

Let $\{W(t)\}$ be a standard Brownian motion with the natural information structure. The independent increments property yields, for $s > t$,

$$E\{W(s) - W(t) \mid W(t)\} = E\{W(s) - W(t)\} = 0.$$

Thus, a standard Brownian motion is a martingale. A Brownian motion with nonzero drift is not a martingale. This is evident from

$$E\{W_{\mu,\sigma}(s) - W_{\mu,\sigma}(t) \mid W_{\mu,\sigma}(t)\} = E\{W_{\mu,\sigma}(s) - W_{\mu,\sigma}(t)\} = \mu(s-t) \neq 0.$$

For a Poisson process $\{N(t), t \geq 0\}$ with intensity λ, it is easy to see

$$\begin{aligned} &E\{[N(s) - \lambda s] - [N(t) - \lambda t] \mid N(t)\} \\ = {} &E\{[N(s) - \lambda s] - [N(t) - \lambda t]\} \\ = {} &E\{N(s) - N(t)\} - \lambda(s - t) = 0. \end{aligned}$$

Thus, $\{N(t) - \lambda t\}$ is a martingale. The term λt is often referred to as the compensator of $\{N(t)\}$.

We now determine when a geometric Brownian motion is a martingale. Let $\{S(t) = S(0)e^{\mu t+\sigma W(t)}\}$. Since $S(t) > 0$, we use the equivalent definition (4.25). For any $s > t$,

$$E\{S(s)/S(t) \mid S(t)\} = E\{S(s)/S(t)\}$$
$$= E\{e^{\mu(s-t)+\sigma(W(s)-W(t))} \mid S(t)\}$$
$$= E\{e^{\mu(s-t)+\sigma(W(s)-W(t))}\} = e^{(\mu+\frac{1}{2}\sigma^2)(s-t)},$$

using the independent increments property of Brownian motion. Thus, $S(t)$ is a martingale if and only if

$$\mu + \frac{1}{2}\sigma^2 = 0. \tag{4.26}$$

Recalling that $\mu + \frac{1}{2}\sigma^2$ is the drift of geometric Brownian motion with parameters μ and σ, (4.26) shows that geometric Brownian motion is a martingale if and only if it has zero drift. As an application, we consider the following example.

Example 4.1   Indifference to Investment Time
Consider a stock whose price follows geometric Brownian motion $\{S(t) = S(0)e^{\mu t+\sigma W(t)}\}$ and a money market account which earns interest at rate r compounded continuously. The discounted value at time 0 of the stock price at time t is thus

$$V(t) = e^{-rt}S(t) = S(0)e^{-rt+\mu t+\sigma W(t)}.$$

To decide when to buy the stock we may look at the expected discounted value $E\{V(t)\}$. For example, for two possible buying times t_1 and t_2, if $E\{V(t_1)\}$ is less than $E\{V(t_2)\}$, we may consider that the stock is cheaper at time t_1 than t_2. Thus the optimal time to purchase the stock is the time that $E\{V(t)\}$ reaches its minimum, and an investor has no preference in terms of investment time if and only if the stochastic process $\{V(t)\}$ is a martingale. As we have shown above, $\{V(t)\}$ is a martingale if and only if

$$-r + \mu + \frac{1}{2}\sigma^2 = 0, \text{ or } r = \mu + \frac{1}{2}\sigma^2.$$

Recalling that $E\{S(t)\} = S(0)e^{(\mu+\frac{1}{2}\sigma^2)t}$, the value $\mu + \frac{1}{2}\sigma^2$ is the rate of return of the stock compounded continuously. Therefore, we may conclude from the analysis above that the investor has no preference in terms of investment time if the return of the stock is equal to the return of the money market. $\quad\square$

In the next example, we present several functions of standard Brownian motion that preserve the martingale structure.

Example 4.2 Functions of Standard Brownian Motion Which Are Martingales
Let $\{W(t)\}$ be standard Brownian motion. Define
(i) $M_1(t) = W(t)^2 - t$:
(ii) $M_2(t) = W(t)^3 - 3tW(t)$;
(iii) $M_3(t) = e^{\lambda W(t) - \frac{1}{2}\lambda^2 t}$.
Each of these stochastic processes is a martingale. It is easy to check that for each fixed t, $E\{|M_i(t)|\}$, $i = 1, 2, 3$, exist. Thus, to verify that these are martingales, we need only examine condition (4.24) or (4.25). For $\{M_1(t)\}$,

$$W(s)^2 - W(t)^2 = [W(s) - W(t)]^2 + 2[W(s) - W(t)]W(t).$$

Thus, for any $s > t$,

$$
\begin{aligned}
&E\left\{ W(s)^2 - W(t)^2 \mid W(t) \right\} \\
= \ &E\left\{ [W(s) - W(t)]^2 \mid W(t) \right\} + 2E\left\{ [W(s) - W(t)]W(t) \mid W(t) \right\}.
\end{aligned}
$$

Since $\{W(t)\}$ is of independent increment,

$$E\left\{ [W(s) - W(t)]^2 \mid W(t) \right\} = E\left\{ [W(s) - W(t)]^2 \right\} = s - t,$$

and

$$
\begin{aligned}
E\{ [W(s) - W(t)]W(t) \mid W(t) \} &= W(t)E\{ W(s) - W(t) \mid W(t) \} \\
&= W(t)E\{ W(s) - W(t) \} = 0.
\end{aligned}
$$

Thus
$$E\left\{ [W(s)^2 - s] - [W(t)^2 - t] \mid W(t) \right\} = 0,$$

i.e., $\{M_1(t)\}$ is a martingale.
For $\{M_2(t)\}$, the derivation is very similar. We use the identity

$$
\begin{aligned}
&[W(s)^3 - 3sW(s)] - [W(t)^3 - 3tW(t)] \\
= \ &[W(s) - W(t)]^3 + 3[W(s) - W(t)]^2 W(t) \\
+ \ &3[W(t)^2 - s][W(s) - W(t)] - 3(s - t)W(t).
\end{aligned}
$$

Since the distribution of $W(s) - W(t)$ is symmetric about 0, all its odd moments are equal to 0. Thus, $E\{ [W(s) - W(t)]^3 \} = 0$. The conditional expectation of the second term is equal to $3(s - t)W(t)$ as shown above and the conditional expectation of the third term is 0. Hence,

$$E\left\{ [W(s)^3 - 3sW(s)] - [W(t)^3 - 3tW(t)] \mid W(t) \right\} = 0,$$

i.e., $\{M_2(t)\}$ is a martingale. $\{M_3(t)\}$ is in fact an alternative expression of geometric Brownian motion. In this form, $\mu = -\frac{1}{2}\lambda^2$ and $\sigma = \lambda$. Thus, (4.26) holds, which implies that $\{M_3(t)\}$ is a martingale.

□

We now consider an important property that characterizes Brownian motion, and is often used to identify whether or not a stochastic process is a Brownian motion.

Martingale Property of Brownian Motion

As we have discussed, a geometric Brownian motion is a martingale if (4.26) holds, i.e. the geometric Brownian motion has zero drift. Let $\{ W_{\mu,\sigma}(t) = \mu t + \sigma W(t) \}$ be Brownian motion with drift μ and volatility σ. Then, for any real number λ, the stochastic process $\{Z_\lambda(t)\}$ given by

$$Z_\lambda(t) = e^{\lambda W_{\mu,\sigma}(t) - \lambda \mu t - \frac{1}{2}\lambda^2 \sigma^2 t}$$

is a martingale for any real λ. This is because

$$\{Z_\lambda(t)\} = e^{\lambda \sigma W(t) - \frac{1}{2}\lambda^2 \sigma^2 t}$$

is geometric Brownian motion with parameters

$$\mu^* = -\frac{1}{2}\lambda^2 \sigma^2, \quad \sigma^* = \lambda \sigma,$$

and we have $\mu^* + \frac{1}{2}(\sigma^*)^2 = 0$.

We now consider the following question. Suppose that there is a stochastic process $\{W_{\mu,\sigma}(t)\}$ such that $E\{ W_{\mu,\sigma}(t) \} = \mu t$ and $Var\{ W_{\mu,\sigma}(t) \} = \sigma^2 t$. Under what conditions, is this process Brownian motion? Obviously, it is necessary that

$$Z_\lambda(t) = e^{\lambda W_{\mu,\sigma}(t) - \lambda \mu t - \frac{1}{2}\lambda^2 \sigma^2 t} \tag{4.27}$$

be a martingale (with respect to the natural information structure generated by $\{W_{\mu,\sigma}(t)\}$), as discussed above. In what follows, we show that the martingale condition (4.27) is also a sufficient condition. In other words, if for a stochastic process $\{W_{\mu,\sigma}(t)\}$, the process $\{Z_\lambda(t)\}$ defined in (4.27) is a martingale, then $\{W_{\mu,\sigma}(t)\}$ is a Brownian motion with drift μ and volatility σ.

Assume now $\{Z_\lambda(t)\}$ is a martingale. Then,

$$E\{ Z_\lambda(t) \} = E\{ Z_\lambda(0) \} = 1.$$

Thus

$$E\left\{ e^{\lambda W_{\mu,\sigma}(t)} \right\} = e^{\lambda \mu t + \frac{1}{2}\lambda^2 \sigma^2 t}.$$

Comparing the above function with the moment generating function of a normal random variable, we conclude that for each t, $W_{\mu,\sigma}(t)$ is normal with mean μt and variance $\sigma^2 t$, i.e. $\{W_{\mu,\sigma}(t)\}$ is a Gaussian process with constant drift and volatility[5]. We next show that $\{W_{\mu,\sigma}(t)\}$ has independent increments. For any $s > t$ and real numbers λ_1 and λ_2,

$$
\begin{aligned}
&E\left\{ e^{\lambda_1[W_{\mu,\sigma}(s)-W_{\mu,\sigma}(t)]+\lambda_2 W_{\mu,\sigma}(t)} \right\} \\
=\ &E\left\{ E\left\{ e^{\lambda_1[W_{\mu,\sigma}(s)-W_{\mu,\sigma}(t)]+\lambda_2 W_{\mu,\sigma}(t)} \mid W_{\mu,\sigma}(y), y \le t \right\} \right\} \\
=\ &E\left\{ E\left\{ e^{\lambda_1[W_{\mu,\sigma}(s)-W_{\mu,\sigma}(t)]} \mid W_{\mu,\sigma}(y), y \le t \right\} e^{\lambda_2 W_{\mu,\sigma}(t)} \right\} \\
=\ &e^{\lambda_1\mu(s-t)+\frac{1}{2}\lambda_1^2\sigma^2(s-t)} E\left(e^{\lambda_2 W_{\mu,\sigma}(t)} \right) \\
=\ &e^{\lambda_1\mu(s-t)+\frac{1}{2}\lambda_1^2\sigma^2(s-t)+\lambda_2\mu t+\frac{1}{2}\lambda_2^2\sigma^2 t} \\
=\ &e^{\lambda_1\mu(s-t)+\frac{1}{2}\lambda_1^2\sigma^2(s-t)} e^{\lambda_2\mu t+\frac{1}{2}\lambda_2^2\sigma^2 t}.
\end{aligned}
\tag{4.28}
$$

Thus, the joint moment generating function of $W_{\mu,\sigma}(s)-W_{\mu,\sigma}(t)$ and $W_{\mu,\sigma}(t)$ can be expressed as the product of a function of λ_1 and a function of λ_2. This implies that random variables $W_{\mu,\sigma}(s) - W_{\mu,\sigma}(t)$ and $W_{\mu,\sigma}(t)$ are independent, and furthermore $e^{\lambda_1\mu(s-t)+\frac{1}{2}\lambda_1^2\sigma^2(s-t)}$ and $e^{\lambda_2\mu t+\frac{1}{2}\lambda_2^2\sigma^2 t}$ are their moment generating functions, respectively (see (2.71) and the comments following it). Thus, it is clear that the stochastic process $\{W_{\mu,\sigma}(t)\}$ has independent increments, and for any fixed pair $s > t$, $W_{\mu,\sigma}(s) - W_{\mu,\sigma}(t)$ is normal with mean $\mu(s-t)$ and variance $\sigma^2(s-t)$. Hence, $\{W_{\mu,\sigma}(t)\}$ is Brownian motion.

The importance of this result is that if one is given a stochastic process $\{W_{\mu,\sigma}(t)\}$ and wants to know whether or not it is Brownian motion, one needs only check whether or not the associated stochastic process $\{Z_\lambda(t)\}$ given in (4.27) is a martingale.

The martingale property (4.27) may also be used to derive martingales related to Brownian motion $\{W(t)\}$.

Consider the derivative of $Z_\lambda(t)$,

$$
Z_\lambda'(t) = \frac{\partial Z_\lambda(t)}{\partial \lambda}
$$

with respect to the parameter λ. Since $\{Z_\lambda(t)\}$ is a martingale,

$$
E\{ Z_\lambda(s) \mid W(t) \} = Z_\lambda(t).
\tag{4.29}
$$

Differentiation of (4.29) with respect to λ yields

$$
E\{ Z_\lambda'(s) \mid W(t) \} = Z_\lambda'(t).
\tag{4.30}
$$

---

[5] A process $\{X(t)\}$ is called a Gaussian process if for each t, $X(t)$ is a normal random variable. However, a Gaussian process may not have independent increments.

Hence for any λ, the stochastic process $\{Z'_\lambda(t)\}$ is also a martingale if $E\{\,|Z'_\lambda(t)|\,\}$ exists for each t. It is obvious that $E\{\,|Z'_\lambda(t)|\,\}$ exists since it is a nice function of the standard Brownian motion $W(t)$. Similarly, the second derivative

$$Z''_\lambda(t) = \frac{\partial^2 Z_\lambda(t)}{\partial \lambda^2}$$

and the third derivative

$$Z'''_\lambda(t) = \frac{\partial^3 Z_\lambda(t)}{\partial \lambda^3}$$

are martingales for any λ. It is easy to see from a simple calculation that

$$
\begin{align}
Z'_\lambda(t) &= [\sigma W(t) - \lambda\sigma^2 t]Z_\lambda(t), & (4.31)\\
Z''_\lambda(t) &= [\sigma W(t) - \lambda\sigma^2 t]Z'_\lambda(t) - \sigma^2 t Z_\lambda(t), & (4.32)
\end{align}
$$

and

$$Z'''_\lambda(t) = [\sigma W(t) - \lambda\sigma^2 t]Z''_\lambda(t) - 2\sigma^2 t Z'_\lambda(t). \qquad (4.33)$$

Letting $\lambda = 0$, we obtain three special martingales:

$$
\begin{align}
Z'_0(t) &= \sigma W(t),\\
Z''_0(t) &= \sigma^2[W(t)^2 - t], \text{ and}\\
Z'''_0(t) &= \sigma^3[W(t)^3 - 3tW(t)].
\end{align}
$$

Thus we have reproduced the martingales in Example 4.2 with $\sigma = 1$. Interested readers may verify that $Z_0^{(4)}(t) = \sigma^4[W(t)^4 - 6tW(t)^2 + 3t^2]$ is also a martingale, where $Z_0^{(4)}(t)$ is the fourth derivative of $Z_\lambda(t)$ with respect to λ at $\lambda = 0$. The above martingales also provide a method to calculate the higher moments of standard Brownian motion. For example, to find $E\{\,[W(t)]^4\,\}$, we may use the identity $E\{\,Z_0^{(4)}(t)\,\} = Z_0^{(4)}(0)$ for it is a martingale. Since for $\sigma = 1$,

$$
\begin{align}
E\{\,Z_0^{(4)}(t)\,\} &= E\{\,[W(t)]^4\,\} - 6tE\{\,W(t)^2\,\} + 3t^2\\
&= E\{\,[W(t)]^4\,\} - 3t^2,
\end{align}
$$

and $Z_0^{(4)}(0) = 0$, we obtain $E\{[W(t)]^4\} = 3t^2$.

4.6   STOPPING TIMES AND THE OPTIONAL SAMPLING THEOREM

As in the discrete-time setting, a stopping time $\mathcal{T}$ in the continuous setting is a nonnegative random variable defined on a probability space $(\Omega, \mathcal{F}, P)$ such that for any $t \geq 0$, the event $\{\mathcal{T} \leq t\}$ belongs to $\mathcal{F}_t$, where $\{\mathcal{F}_t\}$ is the information structure associated with $(\Omega, \mathcal{F}, P)$. As explained in Section 3.6, a stopping time represents a stopping rule in which the decision to stop at a specific time t is based on the information up to time t. All of the stopping time properties of Section 3.6 hold in the continuous-time case. The sum of several stopping times is a stopping time, the maximum of several stopping times is a stopping time, and the minimum of several stopping times is a stopping time. An important class of stopping times are the barrier hitting times. A barrier hitting time is the first time a stochastic process reaches the boundary of a given region. For example, the upper barrier hitting time of a stochastic process $\{X(t)\}$, defined as

$$T_a = \inf\{t; \ X(t) \geq a\}, \tag{4.34}$$

where $a > X(0)$, is a stopping time. Similarly, for $b < X(0)$, the lower barrier hitting time

$$T_b = \inf\{t; \ X(t) \leq b\}$$

is also a stopping time. In the case where $\{X(t)\}$ is a standard Brownian motion, i.e. $X(t) = W(t)$, T_a and T_b are the first passage times discussed in Section 4.3. Their distributions are the one-sided stable distribution of index $1/2$ and are given in (4.10).

Stopping times in financial applications are often in the form of trading times as in the next example.

Example 4.3   A Stop Loss Strategy
ABC Company sells European at-the-money call options on a stock to its clients. In order to limit its exposure to the volatility of the underlying stock, the company implements a stop loss strategy as follows. For each call option written on one share of the stock, buy one share of the stock to cover the call as soon as the stock price is 5% higher than the strike price, and sell it as soon as the stock price is 5% lower than the strike price. Let $S(t)$ denote the stock price at time t. The payoff of a call option at the money maturing at time T is then $\max\{S(T) - K, \ 0\}$, where $K = S(0)$. Define

$$
\begin{aligned}
T_1 &= \inf\{t; \ S(t) \geq (1.05)K\}, \\
T_2 &= \inf\{t; \ T_1 < t, S(t) \leq (0.95)K\}, \\
T_3 &= \inf\{t; \ T_2 < t, S(t) \geq (1.05)K\}, \\
&\quad \cdots
\end{aligned}
$$

$$
\begin{aligned}
T_{2n-1} &= \inf\{t; \ T_{2n-2} < t, S(t) \geq (1.05)K\}, \\
T_{2n} &= \inf\{t; \ T_{2n-1} < t, S(t) \leq (0.95)K\}, \\
&\quad \cdots
\end{aligned}
$$

Then T_{2n-1} represents the n-th time the company buys a share of stock to cover the call option, and T_{2n} represents the n-th time the company sells the stock it holds since the last purchase to take a naked position. Obviously,

$$T_1 < T_2 < \cdots < T_{2n-1} < T_{2n} < \cdots.$$

Thus we obtain a sequence of increasing stopping times. It is easy to see that they are dependent and the correlation among them is complex. For more details on the stop loss strategy, see Hull (1993), Section 14.3.

□

Verification of a continuous stopping time is technically more difficult than verification of a discrete stopping time. One needs to show that $\{T \leq t\}$ belongs to the information structure at time t while in the discrete case we need only show that $\{T = t\}$ belongs to the information structure at time t. However, in practice a random time of our choice is often determined by a stopping rule based on current information. In this case, the random time is a stopping time and there is no need to verify its validity. Nevertheless, we need to proceed with caution.

In what follows, we present the continuous version of the Optional Sampling Theorem. This theorem is very important since it is widely used to calculate barrier hitting time distributions of continuous-time stochastic processes.

The Optional Sampling Theorem (Continuous Version)

If $\{X(t)\}$ is a martingale, then $\{X(T \wedge t), t \geq 0,\}$ is also a martingale. Furthermore, if $|X(T \wedge t)|$ is bounded for all $t \geq 0$, i.e. there is a constant K independent of t such that $|X(T \wedge t)| \leq K, \; t \geq 0$, then

$$E\{X(T)\} = E\{X(0)\}. \tag{4.35}$$

We now apply the Optional Sampling Theorem to calculate some barrier hitting time distributions.

Example 4.4 Upper Barrier Hitting Time of Geometric Brownian Motion Consider geometric Brownian motion

$$S(t) = S(0)e^{\mu t + \sigma W(t)}, \tag{4.36}$$

where $\{W(t)\}$ is standard Brownian motion and $\mu > 0$. For $U > S(0)$, let T_U be the stopping time

$$T_U = \inf\{t, \; S(t) \geq U\}. \tag{4.37}$$

The stopping time T_U represents the first time geometric Brownian motion hits the upper barrier U. As in Example 3.7, we will first identify the Laplace transform of T_U, using the Optional Sampling Theorem.

For any fixed real value $z \geq 0$, let $Z(t) = e^{-zt} [S(t)]^{\xi}$. Then,

$$Z(t) = S(0)^{\xi} e^{(-z+\xi\mu)t+\xi\sigma W(t)}, \tag{4.38}$$

another geometric Brownian motion process. In order to apply the Optional Sampling Theorem one needs to choose $\xi > 0$ such that $\{Z(t)\}$ is a martingale. Since $\{Z(t)\}$ is geometric Brownian motion, it follows from (4.26) that $\{Z(t)\}$ is a martingale if and only if

$$-z + \xi\mu + \frac{1}{2}\xi^2\sigma^2 = 0.$$

Solving this quadratic equation yields the positive solution

$$\xi = \frac{-\mu + \sqrt{\mu^2 + 2\sigma^2 z}}{\sigma^2} = \frac{\mu}{\sigma^2} \left[-1 + \sqrt{1 + 2\left(\frac{\sigma^2}{\mu^2}\right) z} \right].$$

With positive ξ chosen above, the stochastic process $\{Z(t)\}$ is a martingale and the condition of the Optional Sampling Theorem is met (note that if the negative solution is chosen, the condition of the Optional Sampling Theorem is not satisfied, since the process $Z(t) = e^{-zt} [S(t)]^{\xi}$ may be unbounded from below when the value of $S(t)$ approaches zero. This is the reason we require a positive ξ.). Hence,

$$E\{ Z(T_U) \} = Z(0).$$

Equivalently,

$$E\left\{ e^{-zT_U} \right\} = \left(\frac{S(0)}{U} \right)^{\xi} = e^{-a\xi},$$

where $a = \ln\left(\frac{U}{S(0)} \right) > 0$. Therefore,

$$E\left\{ e^{-zT_U} \right\} = e^{a\mu/\sigma^2 \left[1-\sqrt{1+2(\sigma^2/\mu^2)z}\right]}. \tag{4.39}$$

Comparing (4.39) with the Laplace transform of the inverse Gaussian distribution in Example 2.8, the stopping time T_U is inverse Gaussian with shape parameter $a\mu/\sigma^2$ and scale parameter μ^2/σ^2. An immediate implication is that with positive μ, the geometric Brownian motion will hit any upper barrier sooner or later. $\square$

Example 4.5 Lower Barrier Hitting Time of a Geometric Brownian Motion
Consider again the geometric Brownian motion

$$S(t) = S(0)e^{\mu t+\sigma W(t)},$$

which is given in (4.36) of Example 4.4. Let L be such that $S(0) > L$ and T_L be the stopping time

$$T_L = \inf\{t, \ S(t) \le L\}. \tag{4.40}$$

The stopping time T_L represents the first time the geometric Brownian motion hits the lower barrier L. Again, let $Z(t) = e^{-zt} \left[S(t)\right]^\xi$. Then, from Example 4.4, $\{Z(t)\}$ is a martingale if and only if

$$\xi = \frac{\mu}{\sigma^2} \left[-1 \pm \sqrt{1 + 2\left(\frac{\sigma^2}{\mu^2}\right) z} \right].$$

In Example 4.4, we have chosen the positive solution to meet the condition of the Optional Sampling Theorem. With the lower bound L, the process $\{Z(t)\}$ is always bounded from below before the hitting time T_L. Hence, we choose the negative solution

$$\xi = \frac{\mu}{\sigma^2} \left[-1 - \sqrt{1 + 2\left(\frac{\sigma^2}{\mu^2}\right) z} \right]$$

so that $\{Z(t)\}$ is bounded from above. With this choice, the condition of the Optional Sampling Theorem is met and one has

$$E\{ Z(T_L) \} = Z(0).$$

Let $b = \ln \left(\frac{S(0)}{L} \right)$. Then $b > 0$ and

$$E\left\{ e^{-zT_L} \right\} = \left(\frac{S(0)}{L} \right)^\xi = e^{b\xi}.$$

One obtains

$$E\left\{ e^{-zT_L} \right\} = e^{b\mu/\sigma^2 \left[-1 - \sqrt{1 + 2(\sigma^2/\mu^2)z} \right]}. \tag{4.41}$$

Setting $z = 0$ one sees that the value of the right-hand side of (4.41) is $e^{-2b\mu/\sigma^2} < 1$. Thus, the hitting time T_L is not finite, i.e., $P\{ T_L = \infty \} > 0$. To see this, one writes

$$\begin{aligned} E\left\{ e^{-zT_L} \right\} &= E\left\{ e^{-zT_L} \mid T_L < \infty \right\} P\{ T_L < \infty \} \\ &+ E\left\{ e^{-zT_L} \mid T_L = \infty \right\} P\{ T_L = \infty \}. \end{aligned} \tag{4.42}$$

The second term of (4.42) is equal to 0 since the value of the random variable e^{-zT_L} in this case is always 0. Thus

$$E\left\{ e^{-zT_L} \right\} = E\left\{ e^{-zT_L} \mid T_L < \infty \right\} P\{ T_L < \infty \}.$$

Letting $z = 0$ we obtain

$$E\left\{ e^{-zT_L} \right\}\bigg|_{z=0} = P\left\{ T_L < \infty \right\}.$$

Thus $P\left\{ T_L < \infty \right\} = e^{-2b\mu/\sigma^2} < 1$, or $P\left\{ T_L = \infty \right\} = 1 - e^{-2b\mu/\sigma^2} > 0$.
Further, from (4.42), we have

$$E\left\{ e^{-zT_L} \mid T_L < \infty \right\} P\left\{ T_L < \infty \right\}$$

$$= e^{b\mu/\sigma^2\left[-1-\sqrt{1+2(\sigma^2/\mu^2)z}\right]}$$

$$= e^{b\mu/\sigma^2\left[1-\sqrt{1+2(\sigma^2/\mu^2)z}\right]} P\left\{ T_L < \infty \right\}.$$

Hence,

$$E\left\{ e^{-zT_L} \mid T_L < \infty \right\} = e^{b\mu/\sigma^2\left[1-\sqrt{1+2(\sigma^2/\mu^2)z}\right]}. \tag{4.43}$$

It is clear that the right-hand side of (4.43) is the Laplace transform of the inverse Gaussian distribution with shape parameter $b\mu/\sigma^2$ and scale parameter μ^2/σ^2. In summary, the probability that the geometric Brownian motion with positive parameter μ never hits the lower barrier L is $1 - e^{-2b\mu/\sigma^2}$, and given that the geometric Brownian motion process hits the lower barrier L, the hitting time follows an inverse Gaussian distribution with parameters given above.

□

Example 4.6 An Application to a Digital Option
Digital options are options where the payoff function is a step function. In this example, we consider a digital option, called cash-or-nothing option. It pays a notional amount when the price of the underlying asset reaches a specified level during a specified period of time.

Consider a stock whose price follows the geometric Brownian motion

$$S(t) = S(0)e^{\mu t + \sigma W(t)}.$$

Let $U > S(0)$ be a predetermined price level. A cash-or-nothing option written for period $[0, T]$ is such that if the stock reaches the price level U before or at time T, the option writer pays one monetary unit to the option buyer immediately. Otherwise, no payment will be made. To evaluate this option, we calculate the expected discounted payoff when the constant force of interest is r.

Let T_U be the first time that the stock reaches the price level U. Obviously, the discounted payoff as a random variable is $e^{-rT_U}\mathbf{I}_{\{T_U \leq T\}}$, where $\mathbf{I}_{\{T_U \leq T\}}$ is the indicator random variable of the event $\{T_U \leq T\}$. Thus the expected discounted payoff is $E\left\{ e^{-rT_U} \mathbf{I}_{\{T_U \leq T\}} \right\}$. As we have seen in Example 4.4, T_U follows an inverse Gaussian distribution with shape parameter $\alpha = a\mu/\sigma^2$ and scale parameter $\beta = \mu^2/\sigma^2$, where $a = \ln\left(\frac{U}{S(0)}\right)$. Example 2.8 of Chapter 2 shows that the density function of T_U is

$$f_U(t) = \frac{\alpha}{\sqrt{2\pi\beta t^3}} e^{-(\beta t-\alpha)^2/2\beta t} = \frac{a}{\sqrt{2\pi t^3}\sigma} e^{-(\mu t-a)^2/2\sigma^2 t}, \quad t > 0. \quad (4.44)$$

Thus, as in the derivation of the Laplace transform of the inverse Gaussain distribution, one has

$$
\begin{aligned}
& E\left\{ e^{-rT_U} \mathbf{I}_{\{T_U \leq T\}} \right\} \\
= & \int_0^T e^{-rt} f_U(t) dt = \int_0^T e^{-rt} \frac{a}{\sqrt{2\pi t^3}\sigma} e^{-(\mu t-a)^2/2\sigma^2 t} dt \\
= & \int_0^T \frac{a}{\sqrt{2\pi t^3}\sigma} e^{-[(\mu t-a)^2+2r\sigma^2 t^2]/2\sigma^2 t} dt \\
= & e^{(a/\sigma^2)\left[\mu-\sqrt{\mu^2+2r\sigma^2}\right]} \int_0^T \frac{a}{\sqrt{2\pi t^3}\sigma} e^{-\left(t\sqrt{\mu^2+2r\sigma^2}-a\right)^2/2\sigma^2 t} dt.
\end{aligned}
$$

The integrand above is an inverse Gaussian density with parameters $\alpha^* = (a/\sigma^2)\sqrt{\mu^2 + 2r\sigma^2}$ and $\beta^* = (\mu^2 + 2r\sigma^2)/\sigma^2$. It follows from (2.19) that

$$
\begin{aligned}
& \int_0^T \frac{a}{\sqrt{2\pi t^3}\sigma} e^{-\left(t\sqrt{\mu^2+2r\sigma^2}-a\right)^2/2\sigma^2 t} dt \\
= & N\left(\frac{\beta^* T - \alpha^*}{\sqrt{\beta^* T}}\right) + e^{2\alpha^*} N\left(-\frac{\beta^* T + \alpha^*}{\sqrt{\beta^* T}}\right) \\
= & N\left(\frac{\sqrt{\mu^2 + 2r\sigma^2}T - a}{\sigma\sqrt{T}}\right) \\
& + e^{2(a/\sigma^2)\sqrt{\mu^2+2r\sigma^2}} N\left(-\frac{\sqrt{\mu^2 + 2r\sigma^2}T + a}{\sigma\sqrt{T}}\right).
\end{aligned}
$$

Thus

$$
\begin{aligned}
& E\left\{ e^{-rT_U} \mathbf{I}_{\{T_U \leq T\}} \right\} \\
= & e^{(a/\sigma^2)\left[\mu-\sqrt{\mu^2+2r\sigma^2}\right]} N\left(\frac{\sqrt{\mu^2 + 2r\sigma^2}T - a}{\sigma\sqrt{T}}\right) \\
& + e^{(a/\sigma^2)\left[\mu+\sqrt{\mu^2+2r\sigma^2}\right]} N\left(-\frac{\sqrt{\mu^2 + 2r\sigma^2}T + a}{\sigma\sqrt{T}}\right). \quad (4.45)
\end{aligned}
$$

In practice, an option is evaluated under a so called risk-neutral probability measure and the expected discounted payoff under this measure is the time-0 price of the option. If a geometric Brownian motion is assumed for a stock price as described above, then the risk-neutral probability measure corresponds to $\mu = r - \frac{1}{2}\sigma^2$. In this case, we have

$$\sqrt{\mu^2 + 2r\sigma^2} = \sqrt{(r - \frac{1}{2}\sigma^2)^2 + 2r\sigma^2} = r + \frac{1}{2}\sigma^2.$$

Thus, from (4.45) the price of the cash-or-nothing option maturing at time T is

$$\left[\frac{S(0)}{U}\right] N \left(\frac{\left(r + \frac{1}{2}\sigma^2\right) T - \ln\left(U/S(0)\right)}{\sigma\sqrt{T}} \right)$$

$$+ \left[\frac{U}{S(0)}\right]^{\frac{2r}{\sigma^2}} N \left(\frac{-\left(r + \frac{1}{2}\sigma^2\right) T - \ln\left(U/S(0)\right)}{\sigma\sqrt{T}} \right). \qquad (4.46)$$

□

CHAPTER 5

STOCHASTIC CALCULUS: BASIC TOPICS

5.1 STOCHASTIC (ITO) INTEGRATION

The building block of stochastic calculus is stochastic integration with respect to standard Brownian motion[1]. Unlike deterministic calculus which deals with differentiation and integration of deterministic functions, stochastic calculus focuses on integration of stochastic processes. This is due in part to the nondifferentiability of Brownian motion. As we have seen in Chapter 4, the path of Brownian motion takes sharp turns everywhere and thus is nowhere differentiable. As a result, many stochastic processes which are driven by Brownian motion are also nowhere differentiable. The extension of deterministic integration to stochastic integration is not trivial, particularly when the integrand is also a stochastic process. This is because the

---

[1] This is a stochastic counterpart of so-called Riemann-Stieltjes integration in the deterministic world, where we integrate a deterministic function with respect to another deterministic function. See Rudin (1976), Chapter 6.

path of Brownian motion does not have bounded variation[2], a basic requirement for the existence of a Riemann-Stieltjes integral. The complex nature of Brownian motion forces us to approach stochastic integration differently from deterministic integration. Instead of evaluating a stochastic integral on a path-by-path basis[3], we view the Riemann-Stieltjes sums[4] corresponding to the stochastic integral as a sequence of random variables and examine under what condition this sequence will converge (in some sense to be clarified later) to the stochastic integral. However, there are some technical issues that need to be resolved.

In what follows, we lay out the steps in the construction of a stochastic integral, called the Ito integral. There are other types of stochastic integrals but they have little application in financial modelling. Let $\{X(t)\}$ be a stochastic process on $(\Omega, \mathcal{F}, \mathcal{B}_t, P)$, $0 \leq t \leq T$, where $\mathcal{B}_t$ is the natural information structure generated by standard Brownian motion $\{W(t)\}$. We wish to define the Ito integral $\int_a^b X(t)dW(t)$.

Similar to a Riemann-Stieltjes integral in deterministic calculus, for a partition on $[a, b]$, $\mathcal{P} : a = t_0 < t_1 < \cdots < t_{J-1} < t_J = b$, we construct the Riemann-Stieltjes sum

$$S[\mathcal{P}, X] = \sum_{j=1}^{J} X(t_{j-1})[W(t_j) - W(t_{j-1})]. \tag{5.1}$$

It is worth noting that the values of $X(t)$ are taken at the left-end point of each subinterval $[t_{j-1}, t_j]$ and this has central importance in stochastic calculus.

It is clear that $S[\mathcal{P}, X]$ is a random variable on $\mathcal{B}_b$. If $S[\mathcal{P}, X]$ converges to a random variable as $\max |t_j - t_{j-1}| \to 0$, this random variable is then defined as $\int_a^b X(t)dW(t)$, the stochastic integral of $\{X(t)\}$ with respect to $\{W(t)\}$ over $[a, b]$. Two questions arise immediately. What type of convergence is appropriate, and when does the limit of $S[\mathcal{P}, X]$ exist as $\max |t_j - t_{j-1}| \to 0$? In order to answer these questions, an additional assumption on the stochastic process $\{X(t)\}$ is necessary. An assumption that will be used throughout the

---

[2] A deterministic function has bounded variation if it can be expressed as the difference of two nondecreasing functions.

[3] Sometimes, we may treat a stochastic integral as a usual Riemann integral as follows. For a given stochastic process $\{X(t)\}$ and for each outcome ω, $X(t, \omega)$, $a \leq t \leq b$, is a deterministic function over $[a, b]$, where $X(t, \omega)$ is the value of $X(t)$ if ω occurs. Thus, we may evaluate $\int_a^b X(t, \omega)dt$ in the Riemannian sense. When this is done for all ω's, we obtain a random variable that we denote by $\int_a^b X(t)dt$ and call the stochastic integral on a path-by-path basis.

[4] Recall that a Riemann-Stieltjes sum of a deterministic integral $\int_a^b f(t)dg(t)$ has the form $\sum_{j=1}^{J} f(t_{j-1})[g(t_j) - g(t_{j-1})]$, $a = t_0 < t_1 < \cdots < t_J = b$. In the current situation, the functions f and g are replaced by respective stochastic processes.

rest of this book is the square integrability condition

$$E\left\{\int_a^b X^2(t)dt\right\} < \infty, \tag{5.2}$$

where the integral $\int_a^b X^2(t)dt$ is the ordinary Riemann integral on a path-by-path basis. We remark that condition (5.2) is satisfied if the stochastic process $\{X(t)\}$ is bounded and its path is either continuous or piece-wise constant (i.e a step function).

It can be shown that under (5.2), there is a random variable, denoted as $\int_a^b X(t)dW(t)$, such that

$$\lim_{\max|t_j - t_{j-1}| \to 0} E\left\{\left[S[\mathcal{P}, X] - \int_a^b X(t)dW(t)\right]^2\right\} = 0. \tag{5.3}$$

The convergence used in (5.3) is referred to as L^2-convergence. Sometimes, (5.3) is written as

$$\int_a^b X(t)dW(t)$$

$$= \lim_{\max|t_j - t_{j-1}| \to 0} \sum_{j=1}^J X(t_{j-1})[W(t_j) - W(t_{j-1})], \quad \text{in } L^2(\Omega, P).$$

where Ω and P specify the state space and the probability measure where the expectation is taken. The lengthy and complicated proof of (5.3) is omitted. We present only the basic ideas used in the proof. First, assume that $\{X(t)\}$ is a simple process, i.e. there is a partition $a = a_0 < a_1 < \cdots < a_n = b$ with a finite number of random variables $X_0, X_1, \cdots, X_{n-1}$ such that $X(t) = X_j$ for $a_j \le t < a_{j+1}$. It is fairly easy to show (5.3) under this assumption. The next step is to show (5.3) when $\{X(t)\}$ is a bounded continuous process. In this case, one is able to construct a sequence of simple processes which converges to $\{X(t)\}$ in L^2. The result on a simple process is thus transferred to a bounded continuous process. We then extend (5.3) to bounded processes by expressing a bounded process as the limit of a sequence of bounded continuous processes. Finally, (5.3) is proved for all processes satisfying condition (5.2). The approach is the same: expressing a process satisfying condition (5.2) as the limit of a sequence of bounded processes. For a complete proof, we refer to Øksendal (1998), pp. 26-29.

We now consider the simplest class of Ito integrals where the integrands are deterministic functions.

Example 5.1 Suppose that $X(t) = t$. Then

$$\sum_{j=1}^{J} t_{j-1}[W(t_j) - W(t_{j-1})] = \sum_{j=1}^{J} t_{j-1}W(t_j) - \sum_{j=1}^{J} t_{j-1}W(t_{j-1})$$

$$= -\sum_{j=1}^{J-1} W(t_j)(t_j - t_{j-1}) + t_{J-1}W(t_J) - t_0 W(t_0).$$

Thus

$$\int_a^b t\, dW(t)$$

$$= \lim_{\max |t_j - t_{j-1}| \to 0} -\sum_{j=1}^{J-1} W(t_j)(t_j - t_{j-1}) + t_{J-1}W(t_J) - t_0 W(t_0)$$

$$= -\int_a^b W(t)\, dt + bW(b) - aW(a). \tag{5.4}$$

Suppose $X(t) = f(t)$, a deterministic, differentiable function. A similar argument to the above together with the mean value theorem yields

$$\sum_{j=1}^{J} f(t_{J-1})[W(t_j) - W(t_{j-1})]$$

$$= -\sum_{j=1}^{J-1} W(t_j)[f(t_j) - f(t_{j-1})] + f(t_{J-1})W(t_J) - f(t_0)W(t_0)$$

$$= -\sum_{j=1}^{J-1} W(t_j)f'(\xi_j)(t_j - t_{j-1}) + f(t_{J-1})W(t_J) - f(t_0)W(t_0),$$

where ξ_j satisfies $t_{j-1} \le \xi_j \le t_j$ for each j. By taking the limit, we obtain

$$\int_a^b f(t)\, dW(t) = -\int_a^b W(t)f'(t)\, dt + f(b)W(b) - f(a)W(a). \tag{5.5}$$

Two interesting observations are now presented. First, the formula (5.5) is the integration by parts formula in deterministic calculus and the right-hand side of (5.5) involves only an ordinary Riemann integral. Hence, if the integrand of an Ito integral is deterministic, integration by parts applies and the Ito integral may be calculated in terms of a Riemann integral. Second, the Ito integral (5.5) is a normal random variable. This is because the integral term of (5.5) is a Riemann integral of a Gaussian process (i.e. for each fixed t, $W(t)f'(t)$ is a normal random variable) and thus a normal random variable. (5.5) is then a linear combination of normal random variables.

□

Many properties of the deterministic integration carry over to Ito integration because both integrals are constructed in a similar manner. In the following, we list a few important ones, assuming that Ito integrals involved always exist.

1. Ito integration is additive:

$$\int_a^b [X_1(t) + X_2(t)]dW(t) = \int_a^b X_1(t)dW(t) + \int_a^b X_2(t)dW(t);$$

2. Ito integration is additive in terms of integral intervals: for $a < b < c$,

$$\int_a^c X(t)dW(t) = \int_a^b X(t)dW(t) + \int_b^c X(t)dW(t);$$

3. Scalar multiplication is preserved: for any constant k,

$$\int_a^b kX(t)dW(t) = k\int_a^b X(t)dW(t).$$

However, there are fundamental differences between stochastic and deterministic integrations. As mentioned earlier, $\int_a^b X(t)dW(t)$ is a random variable on $(\Omega, \mathcal{B}_b, P)$. Moreover, under the condition

$$E\left\{ \int_a^b X^2(t)dt \right\} < \infty,$$

the mean and variance of $\int_a^b X(t)dW(t)$ exist and are given by

$$E\left\{ \int_a^b X(t)dW(t) \right\} = 0, \tag{5.6}$$

and

$$Var\left\{ \int_a^b X(t)dW(t) \right\} = E\left\{ \left[\int_a^b X(t)dW(t) \right]^2 \right\} = E\left\{ \int_a^b X^2(t)dt \right\}, \tag{5.7}$$

respectively. To see (5.6), we calculate the mean of the corresponding Riemann-Stieltjes sum

$$\sum_{j=1}^J X(t_{j-1})[W(t_j) - W(t_{j-1})].$$

It follows from the Law of Iterated Expectation that

$$E\big\{X(t_{j-1})[W(t_j) - W(t_{j-1})]\big\}$$
$$= E\big\{E\big\{X(t_{j-1})[W(t_j) - W(t_{j-1})] \mid W(t_{j-1})\big\}\big\}$$
$$= E\big\{X(t_{j-1})E\big\{[W(t_j) - W(t_{j-1})] \mid W(t_{j-1})\big\}\big\} = 0.$$

Thus

$$E\left\{\sum_{j=1}^{J} X(t_{j-1})[W(t_j) - W(t_{j-1})]\right\}$$
$$= \sum_{j=1}^{J} E\{X(t_{j-1})[W(t_j) - W(t_{j-1})]\} = 0.$$

One has

$$E\left\{\int_a^b X(t)dW(t)\right\} = E\left\{\lim_{D\to 0} \sum_{j=1}^{J} X(t_{j-1})[W(t_j) - W(t_{j-1})]\right\}$$
$$= \lim_{D\to 0} E\left\{\sum_{j=1}^{J} X(t_{j-1})[W(t_j) - W(t_{j-1})]\right\} = 0,$$

where $D = \max\{|t_j - t_{j-1}|; \text{ for all } j\}$. It should be pointed out that in the above derivation we use the fact that the expectation and limit are interchangeable, which, although true, is not proved here.

To verify statement (5.7), we have

$$E\left\{\left[\int_a^b X(t)dW(t)\right]^2\right\}$$
$$= E\left\{\lim_{D\to 0} \left[\sum_{j=1}^{J} X(t_{j-1})[W(t_j) - W(t_{j-1})]\right]^2\right\}$$
$$= E\left\{\lim_{D\to 0} \sum_{j=1}^{J}\sum_{i=1}^{J} X(t_{j-1})X(t_{i-1})[W(t_j) - W(t_{j-1})]\right.$$
$$\times \; [W(t_i) - W(t_{i-1})]\Big\}$$
$$= \lim_{D\to 0} E\left\{\sum_{j=1}^{J} X^2(t_{j-1})(t_j - t_{j-1})\right\}$$

$$= E\left\{\lim_{D\to 0}\sum_{j=1}^{J} X^2(t_{j-1})(t_j - t_{j-1})\right\} = E\left\{\int_a^b X^2(t)dt\right\}.$$

where $D = \max\{|t_j - t_{j-1}|;$ for all $j\}$. The next to last equality comes from

$$E\left\{X(t_{j-1})X(t_{i-1})[W(t_j) - W(t_{j-1})][W(t_i) - W(t_{i-1})]\right\}$$

$$= \begin{cases} E\left\{X^2(t_{j-1})[W(t_j) - W(t_{j-1})]^2\right\}, & i = j \\ 0, & i < j \end{cases}$$

$$= \begin{cases} E\left\{X^2(t_{j-1})(t_j - t_{j-1})\right\}, & i = j \\ 0, & i < j. \end{cases}$$

Second and more importantly, the point at which the value of $X(t)$ is taken in each subinterval $[t_{j-1}, t_j]$ is critical. Under Ito integration, we always choose the value at the left-end point. Unlike the usual integration in which the choice of points does not affect the integral, different choices of points will lead to different definitions of stochastic integration. For example, if we choose the midpoint $\frac{t_{j-1}+t_j}{2}$ for each subinterval $[t_{j-1}, t_j]$, the limit obtained from the corresponding Riemann-Stieltjes sum is called the *Stratonovich integral* and it is different from the respective Ito integral. Both the Ito integral and Stratonovich integral have many nice properties (in fact, the Stratonovich integral is 'nicer'.), but the Ito integral is the choice for financial applications. This is not only because the Ito integral is mathematically tractable but also because the use of left end-points in a Riemann-Stieltjes sum makes sense in many financial applications. Very often, the process $\{W(t)\}$ or a functional of $\{W(t)\}$ is used to model the price of a risky asset (stock, stock index, interest rate, etc.) over a given period of time, and $X(t)$ represents the trading strategy at time t. Consider now that the entire period is divided into small trading periods $[t_{j-1}, t_j], j = 1, 2, \cdots$. Then a decision for the trading period $[t_{j-1}, t_j]$ can only be made at the beginning of the period, i.e. at time $t = t_{j-1}$. In other words, the Riemann-Stieltjes sum (5.1) is a sensible choice in this situation.

The following example further demontrates that different choices of points may lead to different integral values.

Example 5.2 Consider the integral $\int_a^b W(t)dW(t)$. i.e. $X(t) = W(t)$ with three different choices of points. First, we choose the left-end point of each subinterval of a partition as we define in the Ito integration. Thus

$$\int_a^b W(t)dW(t)$$

$$= \lim_{D \to 0} \sum_{j=1}^{J} W(t_{j-1})[W(t_j) - W(t_{j-1})]$$

$$= \frac{1}{2} \lim_{D \to 0} \sum_{j=1}^{J} \{W^2(t_j) - W^2(t_{j-1}) - [W(t_j) - W(t_{j-1})]^2\}$$

$$= \frac{1}{2}[W^2(b) - W^2(a)] - \frac{1}{2} \lim_{D \to 0} \sum_{j=1}^{J} [W(t_j) - W(t_{j-1})]^2.$$

The sum in the second term is the sum of squares of independent normal random variables. It is easy to see

$$E\Big\{ \sum_{j=1}^{J} [W(t_j) - W(t_{j-1})]^2 \Big\} = \sum_{j=1}^{J} (t_j - t_{j-1}) = b - a.$$

Since

$$Var\Big(\sum_{j=1}^{J} [W(t_j) - W(t_{j-1})]^2 \Big)$$

$$= 2 \sum_{j=1}^{J} (t_j - t_{j-1})^2 \leq 2(b-a) \max |t_j - t_{j-1}|$$

$$= 2(b-a)D,$$

we have

$$Var\Big(\sum_{j=1}^{J} [W(t_j) - W(t_{j-1})]^2 \Big) \to 0, \quad \text{as } D \to 0.$$

Thus the limit of $\sum_{j=1}^{J} [W(t_j) - W(t_{j-1})]^2$ has mean $b - a$ and variance 0, i.e.

$$\lim_{D \to 0} \sum_{j=1}^{J} [W(t_j) - W(t_{j-1})]^2 = b - a.$$

We then obtain

$$\int_a^b W(t)dW(t) = \frac{W^2(b) - W^2(a) - (b-a)}{2}.$$

Now we use the right-end point of each subinterval instead of the left-end point and denote the corresponding integral as $(R) \int_a^b W(t)dW(t)$. A similar argument yields

$$(R) \int_a^b W(t)dW(t)$$

$$= \lim_{D \to 0} \sum_{j=1}^{J} W(t_j)[W(t_j) - W(t_{j-1})]$$

$$= \frac{1}{2} \lim_{D \to 0} \sum_{j=1}^{J} \{W^2(t_j) - W^2(t_{j-1}) + [W(t_j) - W(t_{j-1})]^2\}$$

$$= \frac{1}{2} \left\{ \left[W^2(b) - W^2(a)\right] + \lim_{D \to 0} \sum_{j=1}^{J} [W(t_j) - W(t_{j-1})]^2 \right\}$$

$$= \frac{W^2(b) - W^2(a) + (b - a)}{2},$$

which is different from what we obtained from the Ito integral.

Finally, if we use the midpoints, the resulting integral is the Stratonovich integral. In this case, we have

$$\int_a^b W(t) \circ dW(t)$$

$$= \lim_{D \to 0} \sum_{j=1}^{J} W(\xi_j)[W(t_j) - W(t_{j-1})], \quad \text{where } \xi_j = \frac{t_{j-1} + t_j}{2}$$

$$= \lim_{D \to 0} \sum_{j=1}^{J} W(t_{j-1})[W(t_j) - W(t_{j-1})]$$

$$+ \lim_{D \to 0} \sum_{j=1}^{J} [W(\xi_j) - W(t_{j-1})][W(t_j) - W(t_{j-1})],$$

where $\int_a^b X(t) \circ dW(t)$ is the standard notation for a Stratonovich integral. The limit of the first summation above apparently approaches $\int_a^b W(t)dW(t)$ and thus is equal to $\frac{W^2(b) - W^2(a) - (b-a)}{2}$ as shown earlier. To compute the limit of the second summation, we derive its mean and variance. The mean is given by, utilizing the independence increments of $W(t)$,

$$E \left\{ \sum_{j=1}^{J} [W(\xi_j) - W(t_{j-1})][W(t_j) - W(t_{j-1})] \right\}$$

$$= \sum_{j=1}^{J} E\{ [W(\xi_j) - W(t_{j-1})][W(t_j) - W(t_{j-1})] \}$$

$$= \sum_{j=1}^{J} E\left\{ [W(\xi_j) - W(t_{j-1})]^2 \right\}$$

$$= \sum_{j=1}^{J}(\xi_j - t_{j-1}) = \sum_{j=1}^{J}\frac{1}{2}(t_j - t_{j-1}) = \frac{1}{2}(b - a).$$

Utilizing the independence among its terms, we have

$$Var\left\{ \sum_{j=1}^{J}[W(\xi_j) - W(t_{j-1})][W(t_j) - W(t_{j-1})] \right\}$$

$$= \sum_{j=1}^{J} Var\left\{ [W(\xi_j) - W(t_{j-1})][W(t_j) - W(t_{j-1})] \right\}$$

$$= \sum_{j=1}^{J}\left\{ Var\left\{ [W(\xi_j) - W(t_{j-1})]^2 \right\} \right.$$

$$+ \quad Var\left\{ [W(\xi_j) - W(t_{j-1})][W(t_j) - W(\xi_j)] \right\} \right\}.$$

Since

$$Var\left\{ [W(\xi_j) - W(t_{j-1})]^2 \right\} = 2(\xi_j - t_{j-1})^2 = \frac{1}{2}(t_j - t_{j-1})^2,$$

and

$$Var\left\{ [W(\xi_j) - W(t_{j-1})][W(t_j) - W(\xi_j)] \right\}$$

$$= \quad E\left\{ [W(\xi_j) - W(t_{j-1})]^2[W(t_j) - W(\xi_j)]^2 \right\}$$

$$= \quad E\left\{ [W(\xi_j) - W(t_{j-1})]^2 \right\} E\left\{ [W(t_j) - W(\xi_j)]^2 \right\}$$

$$= \quad (\xi_j - t_{j-1})(t_j - \xi_j) = \frac{1}{4}(t_j - t_{j-1})^2,$$

the variance is given by

$$Var\left\{ \sum_{j=1}^{J}[W(\xi_j) - W(t_{j-1})][W(t_j) - W(t_{j-1})] \right\}$$

$$= \quad \frac{3}{4}\sum_{j=1}^{J}(t_j - t_{j-1})^2$$

$$\leq \quad \frac{3}{4}(b - a) \max|t_j - t_{j-1}| = \frac{3}{4}(b - a)D.$$

Hence,

$$\lim_{D \to 0} Var\left\{ \sum_{j=1}^{J}[W(\xi_j) - W(t_{j-1})][W(t_j) - W(t_{j-1})] \right\} = 0.$$

Similar to the derivation of the Ito integral for $W(t)$, we conclude that the second summation tends to $\frac{1}{2}(b - a)$. Together, we have

$$\int_a^b W(t) \circ dW(t) = \frac{W^2(b) - W^2(a)}{2}.$$

An interesting observation from this result is that unlike the Ito integral, integration by parts from the deterministic calculus applies to this Stratonovich integral. It can be shown that integration by parts from the deterministic calculus also applies to general Stratonovich integrals.

□

5.2 STOCHASTIC DIFFERENTIAL EQUATIONS

This section introduces *stochastic differential equations (SDEs)*. It contains basic assumptions and definitions, existence and uniqueness conditions, and some examples. Stochastic differential equations are widely used in modelling financial securities. This is partly due to their flexibility and partly due to their mathematical tractability.

Generally speaking, a stochastic differential equation is an equation which involves Ito and Riemann integrals of an unknown stochastic process[5] termed the solution of the SDE. Solutions of SDEs form a very large class of stochastic processes. This class includes Brownian motion and geometric Brownian motion introduced in Chapter 4, and many other stochastic processes used in financial modelling.

A stochastic differential equation has the form

$$X(t) = X(0) + \int_0^t \alpha(s, X(s))ds + \int_0^t \sigma(s, X(s))dW(s), \ 0 \le t \le T.$$
(5.8)

In the equation, $\alpha(t, x)$ and $\sigma(t, x)$ are two continuous deterministic functions. $\{X(t)\}$ is a stochastic process on the filtered space $(\Omega, \mathcal{F}, \mathcal{B}_t, P)$ satisfying (5.8), where $\{\mathcal{B}_t\}$ is the natural information structure generated by $\{W(t)\}$. The first integral is a Riemann integral on a path-by-path basis and the second integral is an Ito integral. We call $\{X(t)\}$ the *solution of the stochastic differential equation* (5.8) with initial value $X(0)$ and for convenience, we also call $\{X(t)\}$ an Ito process although the latter is more general[6]. Further, $\alpha(t, X(t))$ and $\sigma(t, X(t))$ are often referred to as the drift and the infinitesimal deviation of the SDE. In finance, $\sigma(t, X(t))$ is also called the volatility of the

---

[5]For this reason, the term 'stochastic differential equation' is somewhat misleading. Stochastic integral equation would be a better description for this type of equation.

[6]Strictly speaking, an Ito process $\{X(t)\}$ is such that it satisfies $X(t) = X(0) + \int_0^t \alpha(s)ds + \int_0^t \sigma(s)dW(s)$, where $\{\alpha(t)\}$ and $\{\sigma(t)\}$ are stochastic processes with respect to $\{\mathcal{B}_t\}$.

stochastic process $\{X(t)\}$. The equation (5.8) is often written in a differential form as follows:

$$dX(t) = \alpha(t, X(t))dt + \sigma(t, X(t))dW(t), \qquad (5.9)$$

with initial condition $X(0)$, or simply

$$dX = \alpha(t, X)dt + \sigma(t, X)dW. \qquad (5.10)$$

The expression (5.9) or (5.10) looks very similar to an ordinary differential equation (ODE) $dx = f(t, x)dt$. However, there is a fundamental difference between SDEs and ODEs. The ODE $dx = f(t, x)dt$ may be written as $\frac{dx}{dt} = f(t, x)$ and thus the differentials dx and dt in the ODE may be viewed as real variables and be moved around the equation. In other words, an ODE is an equation involving a function and its derivatives. Any solution to an ordinary differential equation is thus differentiable. This is not the case when we deal with a SDE. As discussed in Chapter 4, standard Brownian motion $\{W(t)\}$ is nowhere differentiable. Thus, although $dW(t)$ may sometimes be viewed as a differential it can not be treated like its counterpart in the deterministic calculus. Moreover, since the Ito integral $\int_0^t \sigma(s, X(s))dW(s)$ as a function of t is nowhere differentiable in general, $X(t)$ is not differentiable and $dX(t)$ is also not a differential as in the deterministic calculus. Hence, (5.9) or (5.10) is merely a symbolic representation of equation (5.8). Nevertheless, we may sometimes use (5.9) or (5.10) for symbolic computation as we will see in later chapters and for this reason we call dX a *stochastic differential.* But it is very important to remember that $\frac{dW}{dt}$ and $\frac{dX}{dt}$ make no sense at all and should not be used in any way. For notational convenience, from now on we always use equation (5.9) or (5.10) to represent a stochastic differential equation.

The existence and uniqueness of a solution of a stochastic differential equation have theoretical importance. To guarantee that the SDE (5.10) has a unique solution, additional conditions are imposed on the functions $\alpha(t, x)$ and $\sigma(t, x)$. Two commonly used conditions are the Linear Growth Condition and the Lipschitz Condition.

1. **The Linear Growth Condition**
   There is a constant L such that for any $0 \leq t \leq T$ and x

   $$|\alpha(t, x)| + |\sigma(t, x)| \leq L(1 + |x|). \qquad (5.11)$$

2. **The Lipschitz Condition**
   There is a constant L such that for any $0 \leq t \leq T$, and any x_1 and x_2,

   $$|\alpha(t, x_1) - \alpha(t, x_2)| + |\sigma(t, x_1) - \sigma(t, x_2)| \leq L|x_1 - x_2|. \quad (5.12)$$

If $\alpha(t, x)$ and $\sigma(t, x)$ satisfy the Linear Growth Condition (5.11) and the Lipschitz Condition (5.12), the SDE (5.10) has a unique solution $\{X(t)\}$ on

$[0, T]$ with initial value $X(0)$. However, it is worth pointing out that these two conditions are sufficient but not necessary. A solution to (5.10) may uniquely exist even when conditions (5.11) and (5.12) are violated. Hence, we often attempt to solove a SDE without checking the above conditions in practice.

In the following, we present two simplest SDEs and their solutions.

Example 5.3 Consider

$$dX = \mu dt + \sigma dW, \qquad (5.13)$$

where μ and σ are constants. Then equation (5.8) implies that its solution, with initial value $X(0)$, is

$$X(t) = X(0) + \int_0^t \mu ds + \int_0^t \sigma dW(s) = X(0) + \mu t + \sigma W(t). \quad (5.14)$$

Thus, the solution of (5.13) is a Brownian motion process with starting value $X(0)$, drift μ and volatility σ.

$\square$

Example 5.4 Consider

$$dX = \mu dt + \sigma t dW, \qquad (5.15)$$

where μ and σ are constants. Then equation (5.8) implies that its solution, with initial value $X(0)$, is

$$X(t) = X(0) + \int_0^t \mu ds + \int_0^t \sigma s dW(s).$$

It follows from (5.4) that

$$X(t) = X(0) + \mu t + \sigma t W(t) - \sigma \int_0^t W(s)ds. \qquad (5.16)$$

Thus, the solution of (5.15) is a Gaussian process but not a Brownian motion process. For each t, $X(t)$ is a normal random variable with mean and variance given by

$$E\{X(t)\} = X(0) + \mu t, \quad \text{and} \quad Var\{X(t)\} = \sigma^2 t^3/3.$$

$\square$

Unlike a general stochastic process, it is relatively easy to verify that the solution of a SDE is a martingale. It follows from (5.8) that for any $u > t$,

$$E\{X(u) \mid \mathcal{B}_t\}$$

$$
\begin{aligned}
&= X(0) + E\{\int_0^u \alpha(s, X(s))ds \mid \mathcal{B}_t\} + E\{\int_0^u \sigma(s, X(s))dW(s) \mid \mathcal{B}_t\} \\
&= X(0) + \int_0^t \alpha(s, X(s))ds + E\left\{\int_t^u \alpha(s, X(s))ds \mid \mathcal{B}_t\right\} \\
&+ \int_0^t \sigma(s, X(s))dW(s) + E\left\{\int_t^u \sigma(s, X(s))dW(s) \mid \mathcal{B}_t\right\} \\
&= X(t) + E\{\int_t^u \alpha(s, X(s))ds \mid \mathcal{B}_t\}.
\end{aligned}
$$

Hence, if $X(t)$ is a martingale, then

$$
E\{\int_t^u \alpha(s, X(s))ds \mid \mathcal{B}_t\} = 0.
$$

Since u and t are arbitrary,

$$
\alpha(t, X(t)) = 0, \tag{5.17}
$$

for all t. Otherwise, there is a value t such that $\alpha(t, X(t)) \neq 0$. Without the loss of generality, let $\alpha(tX(t)) > 0$. Choose u so close to t such that $\alpha(sX(s)) > 0$, for all $t \leq s \leq u$. Thus, $E\{X(u) \mid \mathcal{B}_t\} > X(t)$, a contradiction. Therefore, a necessary condition for the solution of a SDE to be a martingale is that the SDE has zero drift.

5.3 ONE-DIMENSIONAL ITO'S LEMMA

In this section we present one of the most important results in stochastic calculus: Ito's Lemma. Ito's Lemma may be viewed as a stochastic version of the Chain Rule in the deterministic calculus. Not only is Ito's Lemma a building block for many other important results in stochastic calculus but it also provides a useful tool for solving stochastic differential equations.

Theorem 5.1 One-Dimensional Ito's Lemma
Let $\{X(t)\}$ be a solution of the stochastic differential equation (5.10) and $g(t, x)$ a deterministic function which is continuously differentiable in t and continuously twice differentiable in x. Then the stochastic process $\{g(t, X(t))\}$ is a solution of the following SDE

$$
\begin{aligned}
dg(t, X) &= \left[\frac{\partial g(t, X)}{\partial t} + \alpha(t, X)\frac{\partial g(t, X)}{\partial x} + \frac{1}{2}\sigma^2(t, X)\frac{\partial^2 g(t, X)}{\partial x^2}\right] dt \\
&+ \sigma(t, X)\frac{\partial g(t, X)}{\partial x}dW.
\end{aligned} \tag{5.18}
$$

An alternative expression is

$$dg(t, X) = \left[\frac{\partial g(t, X)}{\partial t} + \frac{1}{2}\sigma^2(t, X)\frac{\partial^2 g(t, X)}{\partial x^2} \right] dt + \frac{\partial g(t, X)}{\partial x} dX, \quad (5.19)$$

since

$$dX = \alpha(t, X)dt + \sigma(t, X)dW.$$

The proof of Ito's Lemma requires advanced mathematics and is omitted. Instead, we take a heuristic approach to derive formula (5.18) or (5.19). First let us temporarily assume that $X(t)$ is deterministic. In this case, the Chain Rule yields

$$dg(t, X) = \frac{\partial g(t, X)}{\partial t}dt + \frac{\partial g(t, X)}{\partial x}dX.$$

Comparing it with the formula (5.19), we see that (5.19) has an extra term

$$\frac{1}{2}\sigma^2(t, X)\frac{\partial^2 g(t, X)}{\partial x^2}dt.$$

To explain this, consider the Taylor expansion of $g(t, X(t))$ at t

$$dg(t, X(t)) = \frac{\partial g(t, X(t))}{\partial t}dt + \frac{\partial g(t, X(t))}{\partial x}dX(t)$$

$$+ \frac{1}{2}\frac{\partial^2 g(t, X(t))}{\partial t^2}[dt]^2 + \frac{\partial^2 g(t, X(t))}{\partial t \partial x}dt dX(t)$$

$$+ \frac{1}{2}\frac{\partial^2 g(t, X(t))}{\partial x^2}[dX(t)]^2 + \text{ higher order terms.} \quad (5.20)$$

Noting that a differential of a function represents the first-order approximation of that function, any term with order higher than the order of dt is small enough to ignore, and for all practical purposes may be considered as zero. As a result, we drop the higher order terms in (5.20). Similarly, The terms with $dt dX(t)$ or $[dt]^2$ are set to zero since their order is also higher than that of dt. The only term needed to be examined in (5.20) is the one with $[dX(t)]^2$. The differential form of $dX(t)$ gives

$$[dX(t)]^2 = [\alpha(t, X)dt + \sigma(t, X)dW(t)]^2$$

$$= \alpha^2(t, X)[dt]^2 + 2\alpha(t, X)\sigma(t, X)dt dW(t) + \sigma^2(t, X)[dW(t)]^2$$

$$= \sigma^2(t, X)[dW(t)]^2.$$

Since

$$E\{[dW(t)]^2\} = E\{[W(t+dt)-W(t)]^2\} = Var\{[W(t+dt)-W(t)]\} = dt,$$

one has $[dW(t)]^2 = dt$. Hence, $[dX(t)]^2 = \sigma^2(t, X)dt$. Thus, by replacing $[dX]^2$ by $\sigma^2(t, X)dt$ in (5.20), we obtain Ito's Lemma (5.19).

As we have seen above, some product rules are used in the derivation of Ito's Lemma. They are $[dt]^2 = 0$, $[dW(t)]^2 = dt$, $dtdX(t) = 0$, and $[dX(t)]^2 = \sigma^2(t, X)dt$. In general, we have the following product rules.

$\times$	dt	$dW(t)$	$dX_1(t)$
dt	0	0	0
$dW(t)$	0	dt	$\sigma_1 dt$
$dX_2(t)$	0	$\sigma_2 dt$	$\sigma_1\sigma_2 dt$

Table 5.1.   The product rules in stochastic calculus.

where

$$
\begin{aligned}
dX_1 &= \alpha_1(t, X_1)dt + \sigma_1(t, X_1)dW, \\
dX_2 &= \alpha_2(t, X_2)dt + \sigma_2(t, X_2)dW.
\end{aligned}
$$

We now consider applications of Ito's Lemma.

Example 5.5   Let $\{S(t)\}$ be a solution of the following SDE

$$dS = \alpha S dt + \sigma S dW, \tag{5.21}$$

where α and σ are constants.

Since (5.21) cannot be solved directly, we look for a transformation of $S(t)$ such that the equation after the transformation is solvable. By ignoring the diffusion term in (5.21), we have $dS = \alpha S dt$, or $d(\ln S) = \alpha dt$. The latter can be solved directly since $\ln S$ does not appear in the right-hand side of the equation. This suggests that a proper transformation would be $g(t, S) = \ln S$. Applying Ito's Lemma, we have

$$d\ln S = \left(\alpha - \frac{1}{2}\sigma^2\right)dt + \sigma dW,$$

and the right-hand side of the equation above does not contain $\ln S$. Thus, this transformation is indeed a proper one, and from Example 5.3 with $X(t) = \ln S(t)$ and $\mu = \alpha - \frac{1}{2}\sigma^2$,

$$\ln S(t) = \ln S(0) + \left(\alpha - \frac{1}{2}\sigma^2\right)t + \sigma W(t).$$

Therefore, the solution of the equation is

$$S(t) = S(0)e^{(\alpha - \frac{1}{2}\sigma^2)t + \sigma W(t)}, \tag{5.22}$$

a geometric Brownian motion process. Noting that if $\mu = \alpha - \frac{1}{2}\sigma^2$, then $\alpha = \mu + \frac{1}{2}\sigma^2$, which is the expected rate of change compounding continuously. This is why we call $\mu + \frac{1}{2}\sigma^2$ the drift of geometric Brownian motion in Chapter 4.

□

In the following example we consider a SDE that is often used to model short term interest rates.

Example 5.6   Consider a stochastic process $\{r(t)\}$ which satisfies the following SDE

$$dr(t) = [\alpha r(t) + \beta(t)]dt + \sigma dW(t), \tag{5.23}$$

where α and σ are constants, and $\beta(t)$ is a known deterministic function.

We will solve this equation for $\{r(t)\}$ using the method of *variation of constants*. The basic idea of the method of variation of constants is the following. To solve a complicated equation, we first consider a reduced version of this equation. This is done normally by setting some of the equation parameters to zero. A solution to the reduced equation then is found. This solution contains an arbitrary constant since no initial condition is given. It is now assumed that the solution of the complicated equation has the same form as the solution to the reduced equation but the constant is replaced by an unknown stochastic process. A relation between the unknown process and the solution to the original equation is thus established, which results in a simple SDE for the unknown process by using Ito's Lemma. If one can solve the simpler SDE, then the desired solution to the original equation is the reduced solution where the constant in the equation is replaced by the solution of the simpler SDE.

For this example, we consider the special case where $\sigma = 0$, and $\beta(t) = 0$ for all t. In this case, the equation (5.23) becomes an ordinary differential equation

$$\frac{dr}{dt} = \alpha r.$$

This ODE has the solution $Ce^{\alpha t}$ with arbitrary constant C. We are now looking for the solution of (5.23) in the form of

$$r(t) = Y(t)e^{\alpha t},$$

for some unknown stochastic process $\{Y(t)\}$. To obtain a SDE for $\{Y(t)\}$, write $Y(t) = r(t)e^{-\alpha t}$, and apply Ito's Lemma to the function $g(t,r) = re^{-\alpha t}$. Since

$$\frac{\partial}{\partial t}g(t,r) = -\alpha g(t,r), \quad \frac{\partial}{\partial r}g(t,r) = e^{-\alpha t}, \quad \text{and} \quad \frac{\partial^2}{\partial r^2}g(t,r) = 0,$$

it follows from Ito's Lemma that

$$\begin{aligned} dY &= -\alpha Y dt + e^{-\alpha t} dr \\ &= e^{-\alpha t}\beta(t)dt + \sigma e^{-\alpha t}dW. \end{aligned}$$

Noting that the right-hand side of the above equation is free of $Y(t)$, this equation can be solved by integrating both sides directly to obtain

$$Y(t) = Y(0) + \int_0^t e^{-\alpha u}\beta(u)du + \sigma \int_0^t e^{-\alpha u}dW(u).$$

To meet the initial condition, we have $Y(0) = r(0)e^{-\alpha 0} = r(0)$. Therefore, by changing $Y(t)$ back to $r(t)e^{-\alpha t}$, one obtains

$$r(t) = r(0)e^{\alpha t} + \int_0^t e^{\alpha(t-u)}\beta(u)du + \sigma \int_0^t e^{\alpha(t-u)}dW(u). \qquad (5.24)$$

Note that the only random term in (5.24) is

$$\sigma \int_0^t e^{\alpha(t-u)}dW(u).$$

Since the integrand is deterministic, it follows from Example 5.1 that the last term in (5.24) is a normal random variable and from (5.6) and (5.7) that it has a mean of 0 and a variance of $\sigma^2 \int_0^t e^{2\alpha(t-u)}du$. Hence for each fixed t, $r(t)$ is a normal random variable with mean

$$r(0)e^{\alpha t} + \int_0^t e^{\alpha(t-u)}\beta(u)du \qquad (5.25)$$

and variance

$$\sigma^2 \int_0^t e^{2\alpha(t-u)}du = \frac{\sigma^2}{2\alpha}\left[e^{2\alpha t} - 1\right]. \qquad (5.26)$$

Therefore the process $\{r(t)\}$ is a Gaussian process but not a Brownian motion process since Property 2 of Brownian motion is not satisfied.

$\square$

The SDE (5.23) in Example 5.6 is widely used to model short term interest rates which we discuss among other interest rate models in the next section.

5.4 CONTINUOUS-TIME INTEREST RATE MODELS

Consider now a continuously trading bond market over the time period $[0, T]$. For all $0 \le t \le s \le T$, let $P(t,s), t \le s$, be the price of a default-free zero-coupon bond at time t that pays one monetary unit at maturity time s and

$\mathcal{B}_t, 0 \leq t \leq T$ is the information structure generated by the bond prices $P(t,s)$, $0 \leq t \leq s \leq T$. The forward rate compounded continuously for time s that is determined at time t, is defined as

$$f(t,s) = -\frac{\partial \ln P(t,s)}{\partial s}. \tag{5.27}$$

When $t = 0$, the function , $f(0,s)$, $s \geq 0$, represents the initial forward rate curve and may be obtained from the current bond market. Obviously, (5.27) can be rewritten as

$$P(t,s) = e^{-\int_t^s f(t,y)dy}. \tag{5.28}$$

The instantaneous interest rate or short rate $r(t)$ at time t is then given by $r(t) = f(t,t)$. Clearly, the forward rate $f(t,s)$, $0 \leq t \leq s$ and the short rate $\{r(t)\}$ are stochastic processes with respect to $\mathcal{B}_t, 0 \leq t \leq T$. To ensure no arbitrage for the bond market[7], there exists a martingale or risk-neutral measure Q such that for all $s \geq 0$, the process

$$V(t,s) = e^{-\int_0^t r(u)du} P(t,s), \ 0 \leq t \leq s, \tag{5.29}$$

is a martingale. Thus we have

$$V(t,s) = E_Q\{V(s,s)|\mathcal{B}_t\},$$

which leads to

$$P(t,s) = E_Q\left\{ e^{-\int_t^s r(u)du} \Big| \mathcal{B}_t \right\}, \tag{5.30}$$

a relationship between a bond price and the short rate.

The Extended Vasicek or Hull-White Model

In what follows we demonstrate the use of the SDE (5.23) to model the short rate process $\{r(t)\}$. Assume that the short rate $\{r(t)\}$ follows the SDE

$$dr(t) = \kappa(\theta(t) - r(t))dt + \sigma dW(t), \tag{5.31}$$

under the risk-neutral measure Q. Here κ is a positive constant and $\theta(t)$ is a deterministic function to be determined by a market condition. Then (5.31) is obtained by reparametrizing the SDE (5.23) where $\alpha = -\kappa$ and $\beta(t) = \kappa\theta(t)$. It follows from (5.24) that

$$r(t) = r(0)e^{-\kappa t} + \kappa \int_0^t e^{-\kappa(t-u)}\theta(u)du + \sigma \int_0^t e^{-\kappa(t-u)}dW(u). \tag{5.32}$$

The bond price $P(t,s)$ can then be solved using (5.30), where $\{\mathcal{B}_t\}$ is generated by standard Brownian motion $\{W(t)\}$. More generally, it follows from

---

[7]The Fundamental Theorem of Asset Pricing stated in Section 3.6 also holds for the continuous case.

the Fundamental Theorem of Asset Pricing that for any $\mathcal{B}_s$-contingent claim $C(s)$ payable at time s, its time-t price $C(t)$ is given by

$$C(t) = E_Q \left\{ e^{-\int_t^s r(u)du} C(s) \,\middle|\, \mathcal{B}_t \right\}, \tag{5.33}$$

Equation (5.31) used to model a short rate process is referred to as the Extended Vasicek model or the Hull-White model in finance literature. If the function $\theta(t) = \theta$, a constant, equation (5.31) reduces to the so called Vasicek model. The parameter $\theta(t)$ represents the long-run average of the short rate, parameter κ the mean-reverting intensity at which the short rate converges to its long-run average as explained below, and parameter σ the volatility of the short rate. When the short rate $r(t)$ is higher than $\theta(t)$, the drift term $\kappa(\theta(t) - r(t))$ becomes negative. Thus, the short rate has a tendency to go down at intensity κ. Similarly, when the short rate $r(t)$ is lower than $\theta(t)$, it has a tendency to go up at intensity κ. This behavior is often called mean-reversion.

Under the model (5.31), a closed form solution for bond prices is obtainable. In fact, we have

$$P(0, s) = e^{A(s) - B(s)r(0)}, \tag{5.34}$$

where

$$A(s) = -\int_0^s \left[1 - e^{-\kappa(s-u)} \right] \theta(u)du + \frac{\sigma^2}{2\kappa^2} \int_0^s \left[1 - e^{-\kappa(s-u)} \right]^2 du \tag{5.35}$$

and

$$B(s) = \frac{1 - e^{-\kappa s}}{\kappa}. \tag{5.36}$$

Obviously, $A(s)$ and $B(s)$ both are deterministic and independent of the short rate.

To derive (5.34), we substitute (5.32) into (5.30) with $t = 0$. Thus one has

$$
\begin{aligned}
P(0, s) &= E_Q \left\{ e^{-\int_0^s r(v)dv} \right\} \\
&= E_Q \left\{ \exp \left\{ -\int_0^s e^{-\kappa v} dv \, r(0) - \kappa \int_0^s \int_0^v e^{-\kappa(v-u)} \theta(u)du\,dv \right.\right. \\
&\qquad \left.\left. - \sigma \int_0^s \int_0^v e^{-\kappa(v-u)} dW(u)dv \right\} \right\}.
\end{aligned}
$$

It is easy to see that

$$B(s) = \int_0^s e^{-\kappa v} dv = \frac{1 - e^{-\kappa s}}{\kappa},$$

and

$$e^{A(s)} = E_Q \left\{ \exp \left\{ -\kappa \int_0^s \int_0^v e^{-\kappa(v-u)} \theta(u)du\,dv \right. \right.$$

$$- \sigma \int_0^s \int_0^v e^{-\kappa(v-u)} dW(u) dv \Bigg\} \Bigg\}$$

Note that interchanging the order of integrals yields

$$-\kappa \int_0^s \int_0^v e^{-\kappa(v-u)} \theta(u) du dv = -\kappa \int_0^s \int_u^s e^{-\kappa(v-u)} \theta(u) dv du$$

$$= \int_0^s \left[-\kappa \int_u^s e^{-\kappa(v-u)} dv \right] \theta(u) du$$

$$= - \int_0^s \left[1 - e^{-\kappa(s-u)} \right] \theta(u) du,$$

and

$$-\sigma \int_0^s \int_0^v e^{-\kappa(v-u)} dW(u) dv = -\sigma \int_0^s \int_u^s e^{-\kappa(v-u)} dv dW(u)$$

$$= -\frac{\sigma}{\kappa} \int_0^s \left[1 - e^{-\kappa(s-u)} \right] dW(u).$$

One thus has

$$e^{A(s)} = \exp\left\{ - \int_0^s \left[1 - e^{-\kappa(s-u)} \right] \theta(u) du \right\}$$

$$\times \ E\left\{ \exp\left\{ -\frac{\sigma}{\kappa} \int_0^s \left[1 - e^{-\kappa(s-u)} \right] dW(u) \right\} \right\}.$$

We remark that interchanging the order of integrals is not allowed in general for Ito integrals but is allowed when the integrand is deterministic. This is because integration by parts applies in this case, as shown in Example 5.1. It follows from Example 5.1 that the integral $-\frac{\sigma}{\kappa} \int_0^s \left[1 - e^{-\kappa(s-u)} \right] dW(u)$ is a normal random variable. From (5.6) and (5.7) this variable has mean 0 and variance $\frac{\sigma^2}{\kappa^2} \int_0^s [1 - e^{-\kappa(s-u)}]^2 du$. Therefore,

$$E_Q\left\{ \exp\left\{ -\frac{\sigma}{\kappa} \int_0^s \left[1 - e^{-\kappa(s-u)} \right] dW(u) \right\} \right\}$$

$$= \exp\left\{ \frac{\sigma^2}{2\kappa^2} \int_0^s \left[1 - e^{-\kappa(s-u)} \right]^2 du \right\},$$

since this expectation may be viewed as the Laplace transform of the normal random variable $-\frac{\sigma}{\kappa} \int_0^s [1 - e^{-\kappa(s-u)}] dW(u)$ at $z = -1$ (see Example 2.7).

The closed form expression for the time-t bond price $P(t, s)$ is obtainable and it has the so-called affine form[8]

$$P(t, s) = e^{A(t,s) - B(s-t)r(t)}, \tag{5.37}$$

[8] Affine interest rate models will be discussed in Chapter 6

where

$$A(t, s) = -\int_t^s \left[1 - e^{-\kappa(s-u)} \right] \theta(u) du + \frac{\sigma^2}{2\kappa^2} \int_t^s \left[1 - e^{-\kappa(s-u)} \right]^2 du.$$
(5.38)

Again, $A(t, s)$ is deterministic and independent of the short rate $r(t)$. The derivation of (5.37) is similar but uses the fact that the short rate process $\{r(t)\}$ is a Markov process. To see that $\{r(t)\}$ is a Markov process, we express $r(v)$, $v \geq t$, by

$$r(v) = r(t)e^{-\kappa(v-t)} + \kappa \int_t^v e^{-\kappa(v-u)}\theta(u)du + \sigma \int_t^v e^{-\kappa(v-u)}dW(u).$$
(5.39)

It is not difficult to verify (5.39) by following the line of proof in Example 5.6. Obviously, there are only two random terms in (5.39): $r(t)e^{-\kappa(v-t)}$ and $\sigma \int_t^v e^{-\kappa(v-u)}dW(u)$. The former is a function of $r(t)$ but not $r(y)$ for any $y < t$. The latter is independent of the underlying information structure at time t since it is an Ito integral starting at t, and therefore is also independent of $r(y)$ for any $y < t$. Thus, $r(v)$ depends only on the value of $r(t)$, and satisfies the Markov property. Therefore, the time-t price of a default-free zero-coupon bond maturing at time s may be written as

$$P(t, s) = E_Q \left\{ e^{-\int_t^s r(u)du} \,\middle|\, \mathcal{B}_t \right\} = E_Q \left\{ e^{-\int_t^s r(u)du} \,\middle|\, r(t) \right\}.$$
(5.40)

Substitution of (5.39) into (5.40) yields (5.37). We leave the details to interested readers. Since $r(t)$ is normally distributed, $P(t, s)$ is a lognormal random variable, i.e. the bond price process $\{P(t, s), 0 \leq t \leq s\}$ is a lognormal process.

Existence of a closed form solution to the bond price is one of the reasons for the popularity of the extended Vasicek model among financial analysts. Another advantage of the extended Vasicek model is that it is able to reproduce the current zero-coupon yield curve. Recall that the function $\theta(t)$ is arbitrary. Thus, the formula (5.34) allows for choosing $\theta(t)$ such that the prices given in (5.34) match the zero-coupon bond prices in the bond market. However, the extended Vasicek model has some drawbacks. One is that the model could generate negative short rates since $r(t)$ is a normal random variable for each t. Moreover, the model is sensitive to its parameters in the sense that a small change in parameter values may result in a large change in bond prices. This could pose a serious problem when frequent calibration of the model is required.

The One-Factor Gaussian Forward Rate Model

Similar to Section 3.7, we may also model forward rates using stochastic differential equations. Assume that under a risk-neutral probability measure the forward rates $f(t, s), t \leq s$, are governed by the following stochastic differential equations:

$$df(t, s) = \alpha(t, s)dt + \sigma(t, s) \, dW(t), \ 0 \leq t \leq s, \quad (5.41)$$

where $W(t)$ is standard Brownian motion, $\alpha(t, s)$ are the term structure of the forward rate drifts and $\sigma(t, s)$ are the term structure of the forward rate volatilities and are deterministic. The model (5.41) is referred to as the one-factor forward rate model as all the forward rates are driven by a single random source $\{W(t)\}$.

It follows from (5.28) that we may rewrite (5.29) as

$$V(t, s) = e^{-\int_0^t r(u)du - \int_t^s f(t,y)dy}.$$

Let $X(t) = \int_0^t r(u)du + \int_t^s f(t, y)dy$. Then

$$
\begin{aligned}
dX(t) &= r(t)dt - f(t, t)dt + \int_t^s df(t, y)dy \\
&= \int_t^s [\alpha(t, y)dt + \sigma(t, y) \, dW(t)]dy \\
&= \left[\int_t^s \alpha(t, y)dy \right] dt + \left[\int_t^s \sigma(t, y)dy \right] dW(t)
\end{aligned}
$$

Applying Ito's Lemma (5.19) to $V(t, s)$ with $g(t, x) = e^{-x}$, we have

$$
\begin{aligned}
& dV(t, s) \\
&= \frac{1}{2} \left[\int_t^s \sigma(t, y)dy \right]^2 V(t, s)dt - V(t, s)dX(t) \\
&= \left\{ \frac{1}{2} \left[\int_t^s \sigma(t, y)dy \right]^2 - \int_t^s \alpha(t, y)dy \right\} V(t, s)dt \\
&\quad - \left[\int_t^s \sigma(t, y)dy \right] V(t, s)dW(t).
\end{aligned}
$$

Since $V(t, s)$ is a martingale under the risk-neutral measure, the necessary condition (5.17) implies that

$$\int_t^s \alpha(t, y)dy = \frac{1}{2} \left[\int_t^s \sigma(t, y)dy \right]^2 ;$$

for all t and s. Differentiating both sides with respect to s yields

$$\alpha(t, s) = \sigma(t, s) \int_t^s \sigma(t, y)dy. \quad (5.42)$$

Hence under the risk-neutral measure, the term structure of drifts for the forward rates in (5.41) is uniquely determined. Since we assume $\sigma(t, s)$ to be deterministic, so is the drift $\alpha(t, s)$. Thus the forward rates $f(t, s)$ for fixed t follow a normal distribution. In other words, the forward rate processes are Gaussian processes. Advantages of using the one-factor Gaussian forward rate model are that it is mathematically tractable, that the price of contingent claims on bonds can be easily implemented and that the model allows for some flexibility of the volatility structure. However, the one-factor forward rate model has its shortcomings too. An obvious one is that all the forward rates and therefore the bond prices are perfectly correlated as they are driven by a single random source.

The stochastic differential equations for bond prices $P(t, s)$ can be derived easily. It follows from Ito's Lemma (5.19) that $P(t, s)$ satisfies the stochastic differential equations:

$$dP(t, s) = r(t)P(t, s)dt - \left[\int_t^s \sigma(t, y)dy\right] P(t, s)dW(t). \qquad (5.43)$$

The class of forward rate models is in general larger than the class of short rate models in the sense that many in the latter can be reproduced from a forward rate model. Write (5.41) as

$$f(t, s) = f(0, s) + \int_0^t \alpha(u, s)du + \int_0^t \sigma(u, s)dW(u),$$

where $\alpha(u, s)$ is given in (5.42). Let $s = t$ and we have

$$r(t) = f(0, t) + \int_0^t \alpha(u, t)du + \int_0^t \sigma(u, t)dW(u). \qquad (5.44)$$

Thus we obtain a SDE for the short rate

$$dr(t) = \alpha(t)dt + \sigma(t, t) \, dW(t), \qquad (5.45)$$

where

$$\alpha(t) = \frac{\partial f(0, t)}{\partial t} + \int_0^t \frac{\partial \alpha(u, t)}{\partial t} du + \int_0^t \frac{\partial \sigma(u, t)}{\partial t} dW(u), \qquad (5.46)$$

assuming all the derivatives in (5.46) exist. With the above formulas, the previously mentioned extended Vasicek model can be reproduced by setting $\sigma(t, s) = \sigma e^{-\kappa(s-t)}$ for some constants $\sigma > 0$ and $\kappa > 0$, as follows. Obviously,

$$\frac{\partial \sigma(u, t)}{\partial t} = -\kappa\sigma(u, t)$$

and

$$\frac{\partial \alpha(u, t)}{\partial t} = -\kappa\alpha(u, t) + [\sigma(u, t)]^2.$$

It follows from (5.44), (5.45) and (5.46) that

$$
dr(t)
$$
$$
= \left\{ \frac{\partial f(0,t)}{\partial t} + \int_0^t [\sigma(u,t)]^2 du - \kappa \int_0^t \alpha(u,t) du \right.
$$
$$
\left. - \kappa \int_0^t \sigma(u,t) dW(u) \right\} dt + \sigma\, dW(t)
$$
$$
= \left\{ \frac{\partial f(0,t)}{\partial t} + \sigma^2 \int_0^t e^{-2\kappa(t-u)} du + \kappa f(0,t) - \kappa r(t) \right\} dt + \sigma\, dW(t)
$$
$$
= \kappa \left\{ f(0,t) + \frac{1}{\kappa}\frac{\partial f(0,t)}{\partial t} + \frac{\sigma^2}{\kappa^2}[1 - e^{-2\kappa t}] - r(t) \right\} dt + \sigma\, dW(t).
$$

Thus, we have

$$
dr(t) = \kappa\,[\theta(t) - r(t)]\, dt + \sigma\, dW(t),
$$

where

$$
\theta(t) = f(0,t) + \frac{1}{\kappa}\frac{\partial f(0,t)}{\partial t} + \frac{\sigma^2}{\kappa^2}[1 - e^{-2\kappa t}].
$$

The extended Vasicek model is reproduced. The continuous version of the Ho-Lee model can also be reproduced by setting $\kappa = 0$. In that case, $\sigma(t,s) = \sigma$.

5.5 THE BLACK-SCHOLES MODEL AND OPTION PRICING FORMULA

In this section, we consider an application of Ito's Lemma to option pricing.

Suppose that there is a market with a money market account and a risky asset that pays no dividends over time period $[0,\,T]$. At time t, $0 \le t \le T$, the value of the money market account is denoted by $B(t)$ with $B(0) = 1$, and the price of the risky asset at time t is denoted as $S(t)$. Assume that the money market account earns interest at a constant rate r, compounded continuously, and the price of the risky asset follows geometric Brownian motion with drift α and volatility σ. That is,

$$
dB = rBdt, \tag{5.47}
$$
$$
dS = \alpha Sdt + \sigma SdW. \tag{5.48}
$$

Consider now a European option on the risky asset $S(t)$. This option pays $\Phi(S(T))$ at time T, where $\Phi(\cdot)$ is the payoff function of the option and is assumed to be continuous with linear growth. In what follows, we price this option using the no-arbitrage pricing technique similar to that discussed in Section 3.6. In order to meet the option obligation at maturity time T, construct a portfolio consisting of the money market account and the risky asset at time 0. During the contract period of the option, the portfolio is continuously rebalanced under a self-financing strategy in the sense that no new money is injected into the portfolio and no money is taken from the portfolio at any time. If the value of the portfolio at maturity time T is the same as the value of the option payoff, then the value of the portfolio at time 0 must be a fair time-0 price of the option. Otherwise there is an arbitrage opportunity. To see this, consider two possible scenarios: the value of the portfolio is higher than the option price and the value of the portfolio is lower than the option price. In the former case, we may take a short position in the portfolio and use the proceeds to purchase the option. The payment received from the option at maturity can then be used to close the short position. Thus, a profit equal to the difference of the time-0 value of the portfolio and the price of the option is made immediately. In the latter case, we sell the option and use part of the proceeds to construct a portfolio against the option. Again, a profit is made. The same argument applies to the entire contract period and therefore the value of the portfolio is equal to the option price at any time during the contract period of the option.

Let $\phi(t, S)$ be the price of the option at time t where the time-t price of the underlying risky asset is S. Also let $\Theta(t, S)$ be the value of the risky asset in the portfolio at time t. The value in the money market account at time t is then $\phi(t, S) - \Theta(t, S)$. The self-financing strategy implies that

$$d\phi(t, S) = [\phi(t, S) - \Theta(t, S)]\frac{dB}{B} + \Theta(t, S)\frac{dS}{S}. \tag{5.49}$$

This is because the right-hand side of the above equation is the capital gain during the period $[t, \ t + dt)$, which has to be fully reinvested under the self-financing strategy. Thus from (5.47) and (5.48),

$$d\phi(t, S) = [r\phi(t, S) + (\alpha - r)\Theta(t, S)]dt + \sigma\Theta(t, S)dW. \tag{5.50}$$

On the other hand, if we assume that $\phi(t, S)$ satisfies the conditions of Ito's Lemma, we have

$$d\phi(t, S) = \left[\frac{\partial}{\partial t}\phi(t, S) + \alpha S\frac{\partial}{\partial S}\phi(t, S) + \frac{1}{2}\sigma^2 S^2\frac{\partial^2}{\partial S^2}\phi(t, S)\right]dt$$

$$+ \ \sigma S\frac{\partial}{\partial S}\phi(t, S)dW. \tag{5.51}$$

Equating the respective coefficients in these two equations yields

$$\Theta(t, S) = S\frac{\partial}{\partial S}\phi(t, S), \tag{5.52}$$

$$r\phi(t, S) + (\alpha - r)\Theta(t, S) = \frac{\partial}{\partial t}\phi(t, S) + \alpha S\frac{\partial}{\partial S}\phi(t, S)$$

$$+ \frac{1}{2}\sigma^2 S^2\frac{\partial^2}{\partial S^2}\phi(t, S). \tag{5.53}$$

Replacing $\Theta(t, S)$ in (5.53) by the right-hand side of (5.52) yields

$$\frac{\partial}{\partial t}\phi(t, S) + \frac{1}{2}\sigma^2 S^2 \frac{\partial^2}{\partial S^2}\phi(t, S) + rS\frac{\partial}{\partial S}\phi(t, S) - r\phi(t, S) = 0, \quad (5.54)$$

for any $0 \le t \le T$ and $S \ge 0$. Thus, $\phi(t, S)$ is the solution of the partial differential equation (5.54) with terminal condition $\phi(T, S) = \Phi(S)$. The equation (5.54) is called the Black-Scholes Partial Differential Equation.

Readers can verify that the solution of (5.54)[9] is

$$\phi(t, S) = \frac{e^{-r(T-t)}}{\sqrt{2\pi}} \int_{-\infty}^{\infty} \Phi(Se^{(r-\frac{1}{2}\sigma^2)(T-t)+\sigma\sqrt{T-t}\,y})e^{-\frac{y^2}{2}}\,dy. \quad (5.55)$$

The above formula may be interpreted as follows. Consider that there is another probability measure Q on the same state space. Under this new probability measure, the asset price $S(t)$ is still a geometric Brownian motion process but its expected return is exactly the same as the return of the money market account, that is,

$$S(t) = S(0)e^{(r-\frac{1}{2}\sigma^2)t+\sigma W_Q(t)}, \quad (5.56)$$

or it can be expressed in terms of the SDE

$$dS(t) = rS(t)dt + \sigma S(t)dW_Q(t), \quad (5.57)$$

where $W_Q(t), t \ge 0$, is standard Brownian motion under the Q-measure. The option price given in (5.55) can then be written as

$$\phi(t, S) = e^{-r(T-t)}E_Q\Big\{\Phi(S(T)) \mid S(t) = S\Big\}, \quad (5.58)$$

where $E_Q(\cdot)$ is the expectation under the probability measure Q. Thus, formula (5.58) states that the option price at time t is the expected discounted payoff of the option under the probability measure Q under which the underlying risky asset has the same expected return as the money market account. For this reason, the probability measure Q is referred to as the risk-neutral measure. An interesting observation from (5.58) is that the option price is a function of the interest rate r and the asset volatility σ, but not the asset drift α. The procedure of using a self-financing strategy to replicate the payoff of an option or other derivative securities as above is called *delta hedging*.

We now use formula (5.58) or equivalently (5.55) to derive the Black-Scholes formula for European call options. Consider a European call option written on the risky asset $\{S(t)\}$, maturing at time T with strike price K. The payoff of this option is

$$\Phi(S(T)) = \max\{S(T) - K, 0\}.$$

[9]How to solve this equation will be discussed in Section 6.2.

Let $\phi_c(t, S)$ be the price. Then, from (5.58)

$$
\begin{aligned}
&\phi_c(t, S) \\
&= \frac{e^{-r(T-t)}}{\sqrt{2\pi}} \int_{-\infty}^{\infty} \max(S e^{(r-\frac{1}{2}\sigma^2)(T-t)+\sigma\sqrt{T-t}\,y} - K, 0) e^{-\frac{y^2}{2}} dy \\
&= \frac{e^{-r(T-t)}}{\sqrt{2\pi}} \int_{-\infty}^{\infty} \max(S e^{(r-\frac{1}{2}\sigma^2)(T-t)-\sigma\sqrt{T-t}\,x} - K, 0) e^{-\frac{x^2}{2}} dx \\
&= \frac{e^{-r(T-t)}}{\sqrt{2\pi}} \int_{-\infty}^{\frac{\ln(S/K)+(r-\frac{1}{2}\sigma^2)(T-t)}{\sigma\sqrt{T-t}}} (S e^{(r-\frac{1}{2}\sigma^2)(T-t)-\sigma\sqrt{T-t}\,x} - K) e^{-\frac{x^2}{2}} dx \\
&= \frac{S}{\sqrt{2\pi}} \int_{-\infty}^{\frac{\ln(S/K)+(r+\frac{1}{2}\sigma^2)(T-t)}{\sigma\sqrt{T-t}}} e^{-\frac{1}{2}x^2} dx \\
&\quad - \frac{e^{-r(T-t)} K}{\sqrt{2\pi}} \int_{-\infty}^{\frac{\ln(S/K)+(r-\frac{1}{2}\sigma^2)(T-t)}{\sigma\sqrt{T-t}}} e^{-\frac{1}{2}x^2} dx.
\end{aligned}
$$

That is,

$$
\phi_c(t, S) = SN(d_1(t)) - Ke^{-r(T-t)} N(d_2(t)), \tag{5.59}
$$

where

$$
d_1(t) = \frac{\log(S/K) + (r + \frac{1}{2}\sigma^2)(T - t)}{\sigma\sqrt{T - t}} \tag{5.60}
$$

and

$$
d_2(t) = \frac{\log(S/K) + (r - \frac{1}{2}\sigma^2)(T - t)}{\sigma\sqrt{T - t}} = d_1(t) - \sigma\sqrt{T - t}. \tag{5.61}
$$

$N(x)$ is the distribution function of the standard normal random variable. We obtain the well-known Black-Scholes option pricing formula. This formula not only gives the price of the call option but also provides a trading strategy when constructing a portfolio against the option. It can be shown with a little algebra that $\frac{\partial \phi_c(t,S)}{\partial S} = N(d_1)$, i.e $N(d_1)$ is the delta of the option price. This suggests that to construct a portfolio to meet the option obligation, one retains $N(d_1)$ shares of the risky asset and takes a short position in the money market account. The exact amount in the short position is $Ke^{-r(T-t)} N(d_2)$.

The price of a European put option can be obtained easily using the put-call parity

$$
\phi_p(t, S) = \phi_c(t, S) - S + Ke^{-r(T-t)},
$$

and it is

$$
\phi_c(t, S) = -SN(-d_1(t)) + Ke^{-r(T-t)} N(-d_2(t)). \tag{5.62}
$$

The derivation of the formula (5.59) is essentially the calculation of the expectation $E\left\{\max[e^X - K, 0]\right\}$, where X is a normal random variable with mean μ_X and variance σ_X^2. Following the line of the derivation of (5.59), we have

$$E\left\{\max[e^X - K, 0]\right\}$$
$$= e^{\mu_X + \frac{1}{2}\sigma_X^2} N\left(\frac{\mu_X - \ln K + \sigma_X^2}{\sigma_X}\right) - KN\left(\frac{\mu_X - \ln K}{\sigma_X}\right).$$
$$(5.63)$$

The formula (5.63) is sometime convenient to use when pricing a call option on a risky asset that follows a lognormal process.

A slight generalization of (5.63) is also very useful. Consider the expectation

$$E\left\{e^Y \max[e^X - K, 0]\right\}, \qquad (5.64)$$

where X and Y are correlated normal random variables with means μ_X and μ_Y, variances σ_X^2 and σ_Y^2, and covariance σ_{XY}.

To solve (5.64), we let

$$Z = \frac{e^Y}{E\{e^Y\}}.$$

It follows from (3.42) in Chapter 3 that the associated Esscher probability measure with Z is defined as

$$P_Z(A) = E\{\mathbf{I}_A Z\}.$$

Under the Esscher measure, the moment generating function of (X, Y) may be computed as follows.

$$E_Z\{e^{z_1 X + z_2 Y}\} = \frac{E\{e^{z_1 X + z_2 Y} e^Y\}}{E\{e^Y\}}$$
$$= e^{-\mu_Y - \frac{1}{2}\sigma_Y^2} E\{e^{z_1 X + (z_2 + 1)Y}\},$$

since $E\{e^Y\} = e^{\mu_Y + \frac{1}{2}\sigma_Y^2}$ as shown in Example 2.7. Further, the formula (2.78) in Example 2.17 gives

$$E_Z\{e^{z_1 X + z_2 Y}\}$$
$$= e^{-\mu_Y - \frac{1}{2}\sigma_Y^2} E\{e^{z_1 X + (z_2 + 1)Y}\}$$
$$= e^{-\mu_Y - \frac{1}{2}\sigma_Y^2} e^{\mu_X z_1 + \mu_Y(z_2 + 1) + \frac{1}{2}\left(\sigma_X^2 z_1^2 + 2\sigma_{XY} z_1(z_2 + 1) + \sigma_Y^2(z_2 + 1)^2\right)}$$
$$= e^{(\mu_X + \sigma_{XY})z_1 + (\mu_Y + \sigma_Y^2)z_2 + \frac{1}{2}\left(\sigma_X^2 z_1^2 + 2\sigma_{XY} z_1 z_2 + \sigma_Y^2 z_2^2\right)}.$$

Compared the above with (2.78), it is easy to see that under the Esscher measure, (X, Y) is also a bivariate normal with new means $\mu_X + \sigma_{XY}$ and $\mu_Y + \sigma_Y^2$. The variances and the covariance remain the same.

We now ready to solve (5.64).

$$
\begin{aligned}
& E\left\{e^Y \max[e^X - K,\, 0]\right\} \\
= \; & E\{e^Y\} E_Z\{\max[e^X - K,\, 0]\} \\
= \; & e^{\mu_Y + \frac{1}{2}\sigma_Y^2} \left\{ e^{(\mu_X + \sigma_{XY}) + \frac{1}{2}\sigma_X^2} N\left(\frac{(\mu_X + \sigma_{XY}) - \ln K + \sigma_X^2}{\sigma_X} \right) \right. \\
& \quad - KN\left(\frac{(\mu_X + \sigma_{XY}) - \ln K}{\sigma_X} \right) \Bigg\} \\
= \; & e^{\mu_X + \mu_Y + \frac{1}{2}\left(\sigma_X^2 + 2\sigma_{XY} + \sigma_Y^2\right)} N\left(\frac{\mu_X + \sigma_{XY} + \sigma_X^2 - \ln K}{\sigma_X} \right) \\
& \quad - K\, e^{\mu_Y + \frac{1}{2}\sigma_Y^2} N\left(\frac{\mu_X + \sigma_{XY} - \ln K}{\sigma_X} \right).
\end{aligned} \tag{5.65}
$$

Black's 76 Formula

Black's Formula is often used to price call options on a forward or futures contract. A forward contract entered at time t with delivery time T on a risky asset (a finanical instrument or a physical commodity) is an agreement between two parties under which one party (the holder) will buy from the other (the underwriter) certain units of the risky asset at time T for a price determined at time t. The pre-determined price is called the forward price of the contract. As an agreeemnt, there is no cost to both parties at time t but the delivery is obliged by both parties at time T. A futures contract is very similar to a forward contract except that it can be traded in the futures market. As a result, the price of a futures contract changes over the duration of the contract and the positions of the both parties are adjusted accordingly, meaning that gains and loss resulting from the price fluctuation are deducted or credited to the holder's account typically on a daily basis.

Suppose that a forward contract or a futures contract with delivery time T is written on a risky asset $S(t)$, $t \geq 0$, which is governed by the Black-Scholes model (5.48). It can be shown (Björk, 1998) that the theoretical prices of the forward contract and the futures contract based on the risk-neutral valuation coincide. Let $F(t, T)$ be the futures (or forward) price at time t. Zero cost to both parties implies that the time-t value of the payment $F(t, T)$ from the holder is the same as the time-t value of $S(T)$. It follows from (5.48) that

$$
\begin{aligned}
e^{-r(T-t)} F(t, T) &= e^{-r(T-t)} E_Q\{F(t, T) \mid S(t)\} \\
&= e^{-r(T-t)} E_Q\{S(T) \mid S(t)\} = S(t).
\end{aligned}
$$

Hence

$$F(t, T) = e^{r(T-t)} S(t). \tag{5.66}$$

Consider now a European call option on the futures price with strike price K and expiration time s, $t < s < T$. The holder of the option will receive a payment equal to $\max[F(s, T) - K, 0]$ and obtain a long position in the futures contract at time s. Since the long position in the futures contract has zero value, we only need to price the payoff $\max[F(s, T) - K, 0]$. By using (5.48) and (5.66), the price of the futures option is

$$
\begin{aligned}
&\phi_f(t, S) \\
&= e^{-r(s-t)} E_Q\{\max[F(s, T) - K, 0] \mid S(t) = S\} \\
&= e^{-r(s-t)} E_Q\left\{\max\left[e^{r(T-s)} S(s) - K, 0\right] \mid S(t) = S\right\} \\
&= e^{r(T-s)} e^{-r(s-t)} E_Q\left\{\max\left[S(s) - e^{-r(T-s)} K, 0\right] \mid S(t) = S\right\}.
\end{aligned}
$$

Applying (5.59) or (5.63) and (5.66), we have

$$
\begin{aligned}
\phi_f(t, S) &= e^{r(T-s)} S N(\bar{d}_1(t)) - e^{-r(s-t)} K N(\bar{d}_2(t)) \\
&= e^{-r(s-t)} \left[F(t, T) N(\bar{d}_1(t)) - K N(\bar{d}_2(t))\right], \tag{5.67}
\end{aligned}
$$

where

$$\bar{d}_1(t) = \frac{\ln(F(t, T)/K) + \frac{1}{2}\sigma^2(s - t)}{\sigma\sqrt{s - t}}$$

and

$$\bar{d}_2(t) = \bar{d}_1(t) - \sigma\sqrt{s - t}.$$

The formula (5.67) is the well known Black's Formula. Using the put-call parity, a European put option on the futures contract can be derived too. Black's Formula also has applications in pricing interest rate caps and floors.

Finally in this section, we consider a slight generalization of the Black-Scholes model. Instead of assuming a constant volatility σ, we assume that the volatility of the risky asset is a deterministic function $\sigma(t)$. That is, under the Q-measure, the risky asset $S(t)$ follows the SDE

$$dS(t) = rS(t)dt + \sigma(t)S(t)dW_Q(t). \tag{5.68}$$

This model is referred to as Merton's model and it provides some although limited flexibilities for calibration of certain volatility patterns such as volatility smile and volatility skew.

Applying Ito's Lemma with $g(t, S) = \ln S$, we can solve the SDE (5.68) and obtain

$$S(t) = S(0)e^{\int_0^t [r - \frac{1}{2}\sigma^2(y)]dy + \int_0^t \sigma(y)dW_Q(y)}. \tag{5.69}$$

Since $\sigma(t)$ is deterministic, Example 5.1 implies that $S(t)$ is a lognormal process. If we write $S(t) = e^{X(t)}$, then $X(t)$ is normal with mean

$$\ln S(0) + \int_0^t \left[r - \frac{1}{2}\sigma^2(y) \right] dy$$

and variance

$$\int_0^t \sigma^2(y) dy.$$

Furthermore, for any $T > t$, the conditional distribution of $X(T)$ at time t is normal with mean

$$\mu_t = \ln S(t) + \int_t^T [r - \frac{1}{2}\sigma^2(y)] dy$$

and variance

$$\sigma_t^2 = \int_t^T \sigma^2(y) dy. \tag{5.70}$$

A European call option on the risky asset with strike price K and expiration time T now can be valuated easily using (5.66). The time-t price of the option is

$$
\begin{aligned}
&\phi_c(t, S) \\
=\ & e^{-r(T-t)} E_Q \left\{ \max[S(T) - K, \, 0] \mid S(t) = S \right\} \\
=\ & e^{-r(T-t)} \left[e^{\mu_t + \frac{1}{2}\sigma_t^2} N\left(\frac{\mu_t - \ln K + \sigma_t^2}{\sigma_t} \right) - KN\left(\frac{\mu_t - \ln K}{\sigma_t} \right) \right] \\
=\ & SN(d_1(t)) - Ke^{-r(T-t)} N(d_2(t)), \tag{5.71}
\end{aligned}
$$

where

$$d_1(t) = \frac{\ln(S/K) + r(T - t) + \frac{1}{2}\sigma_t^2}{\sigma_t}$$

and

$$d_2(t) = d_1(t) - \sigma_t.$$

5.6 THE STOCHASTIC VERSION OF INTEGRATION BY PARTS

As discussed in earlier sections, the one-dimensional Ito's Lemma allows for identification of the drift and diffusion (volatility) terms of a function of an Ito process and thus provides a very useful tool in solving some SDEs. However, if a stochastic process is a function of two or more Ito processes, the one-dimensional Ito's Lemma is not applicable. Although this problem can be solved using the multi-dimensional Ito's Lemma, we instead consider in this section a simpler but useful case where a stochastic process is a product of two Ito processes. The multi-dimensional Ito's Lemma will be presented in Chapter 6.

Theorem 5.2 Integration by Parts

Let $\{X_1(t)\}$ and $\{X_2(t)\}$ be solutions of the following SDEs:

$$dX_1 = \alpha_1(t, X_1)dt + \sigma_1(t, X_1)dW, \tag{5.72}$$

and

$$dX_2 = \alpha_2(t, X_2)dt + \sigma_2(t, X_2)dW, \tag{5.73}$$

respectively. Then,

$$d(X_1 X_2) = X_2 dX_1 + X_1 dX_2 + \sigma_1(t, X_1)\sigma_2(t, X_2)dt \tag{5.74}$$

or

$$\begin{aligned} d(X_1 X_2) &= [\alpha_1(t, X_1)X_2 + \alpha_2(t, X_2)X_1 + \sigma_1(t, X_1)\sigma_2(t, X_2)]dt \\ &+ [\sigma_1(t, X_1)X_2 + \sigma_2(t, X_2)X_1]dW. \end{aligned} \tag{5.75}$$

Again, we omit the mathematical proof and proceed with a heuristic derivation of the above fomula using the product rules given in Section 5.3. Write a stochastic differential $dX(t)$ as $X(t + dt) - X(t)$. Thus one has

$$\begin{aligned} d(X_1 X_2) &= X_1(t + dt)X_2(t + dt) - X_1(t)X_2(t) \\ &= X_2(t)[X_1(t + dt) - X_1(t)] + X_1(t)[X_2(t + dt) - X_2(t)] \\ &+ [X_1(t + dt) - X_1(t)][X_2(t + dt) - X_2(t)] \\ &= X_2(t)dX_1(t) + X_1(t)dX_2(t) + dX_1(t)dX_2(t). \end{aligned}$$

It follows from the product rules that $dX_1(t)dX_2(t) = \sigma_1(t, X_1)\sigma_2(t, X_2)dt$. Thus the formula (5.74) follows immediately.

Ito's Lemma and integration by parts are very powerful tools, especially when they are used together, as shown in many results of this and later sections.

We first drive the SDE for the ratio of the two Ito processes $X_1(t)/X_2(t)$. The ratio of two Ito processes occurs commonly in finnancial modelling as many discounted value processes are of this form. Assume that $X_2(t) > 0$ (or $X_2(t) < 0$) for all t. Applying Ito's Lemma (5.19) to $X_2(t)$ with $g(t, x) = 1/x$, we have

$$d\frac{1}{X_2} = \frac{\sigma_2^2(t, X_2)}{X_2^3}dt - \frac{1}{X_2^2}dX_2.$$

Thus

$$d\frac{X_1}{X_2} = \frac{1}{X_2}dX_1 + X_1 d\frac{1}{X_2} - \frac{\sigma_1(t, X_1)\sigma_2(t, X_2)}{X_2^2}dt$$

$$= \frac{1}{X_2}dX_1 - \frac{X_1}{X_2^2}dX_2 + \left[\frac{\sigma_2^2(t, X_2)X_1}{X_2^3} - \frac{\sigma_1(t, X_1)\sigma_2(t, X_2)}{X_2^2} \right] dt.$$

$$(5.76)$$

In the next example, we solve a linear SDE again using Ito's Lemma and integration by parts.

Example 5.7 Consider the following SDE

$$dX = (\alpha X + \beta)dt + (\gamma + \sigma X)dW, \qquad (5.77)$$

where α, β, γ and σ are constants. We will show how to apply integration by parts to solve this equation. The basic approach is the method of variation of constants as discussed in Example 5.6.

First let us assume $\gamma = 0$, i.e.

$$dX = (\alpha X + \beta)dt + \sigma X dW. \qquad (5.78)$$

Apply the method of variation of constants to (5.78) by letting $\beta = 0$. Thus, we obtain the SDE in Example 5.5 and have a general solution $C\Psi(t)$, where C is an arbitrary constant and $\Psi(t)$ is geometric Brownian motion

$$\Psi(t) = e^{(\alpha - \frac{1}{2}\sigma^2)t + \sigma W(t)}.$$

As in Example 5.6, we look for a solution of the form $X(t) = Y(t)\Psi(t)$, for an unknown stochastic process $\{Y(t)\}$. Since $Y(t) = \Psi^{-1}(t)X(t)$ where $\Psi^{-1}(t)$ is the reciprocal of $\Psi(t)$, we may apply integration by parts as long as we can express $\Psi^{-1}(t)$ as a solution of a SDE. Since

$$\Psi^{-1}(t) = e^{-(\alpha - \frac{1}{2}\sigma^2)t - \sigma W(t)},$$

it is geometric Brownian motion with drift $-(\alpha - \frac{1}{2}\sigma^2) + \frac{1}{2}\sigma^2 = -\alpha + \sigma^2$ and volatility $-\sigma$. Hence, it satisfies

$$d\Psi^{-1} = (-\alpha + \sigma^2)\Psi^{-1}dt - \sigma\Psi^{-1}dW.$$

By Theorem 5.2,

$$\begin{aligned} dY &= X d\Psi^{-1} + \Psi^{-1}dX - \sigma^2\Psi^{-1}X dt \\ &= (-\alpha + \sigma^2)\Psi^{-1}X dt - \sigma\Psi^{-1}X dW \\ &+ \alpha\Psi^{-1}X dt + \beta\Psi^{-1}dt + \sigma\Psi^{-1}X dW - \sigma^2\Psi^{-1}X dt \\ &= \beta\Psi^{-1}dt. \end{aligned}$$

Noting that the right-hand side of the equation is free of $Y(t)$, one has

$$Y(t) = Y(0) + \beta \int_0^t \Psi^{-1}(s)ds = Y(0) + \beta \int_0^t e^{-(\alpha - \frac{1}{2}\sigma^2)s - \sigma W(s)}ds.$$

Therefore,

$$
\begin{aligned}
X(t) &= [Y(0) + \beta \int_0^t e^{-(\alpha-\frac{1}{2}\sigma^2)s-\sigma W(s)}ds]\Psi(t) \\
&= X(0)e^{(\alpha-\frac{1}{2}\sigma^2)t+\sigma W(t)} + \beta \int_0^t e^{(\alpha-\frac{1}{2}\sigma^2)(t-s)+\sigma[W(t)-W(s)]}ds.
\end{aligned}
$$

$$(5.79)$$

Return now to the case $\gamma \neq 0$. Let $\tilde{X}(t) = X(t) + \frac{\gamma}{\sigma}$, where $X(t)$ is the solution to (5.77). Then

$$
d\tilde{X} = \left(\alpha \tilde{X} + \beta - \frac{\alpha\gamma}{\sigma}\right)dt + \sigma \tilde{X}dW,
$$

i.e. $\tilde{X}(t)$ satisfies (5.79) with β replaced by $\beta - \frac{\alpha\gamma}{\sigma}$. Thus, we have

$$
\begin{aligned}
X(t) &= \tilde{X}(t) - \frac{\gamma}{\sigma} \\
&= -\frac{\gamma}{\sigma} + \left[X(0) + \frac{\gamma}{\sigma}\right]e^{(\alpha-\frac{1}{2}\sigma^2)t+\sigma W(t)} \\
&+ \left(\beta - \frac{\alpha\gamma}{\sigma}\right)\int_0^t e^{(\alpha-\frac{1}{2}\sigma^2)(t-s)+\sigma[W(t)-W(s)]}ds. \quad (5.80)
\end{aligned}
$$

$\square$

5.7   EXPONENTIAL MARTINGALES

Positive martingales play a central role in changing probability measures. As we have seen in Section 3.5, to obtain desirable properties for a discrete-time stochastic process, one needs to change the underlying probability measure to an equivalent probability measure. To accomplish this, we usually must find a positive martingale with unit starting value. This is also true for continuous-time stochastic processes. Moreover, many continuous positive martingales used in the change of measures have an exponential form in connection with Ito processes, as will be seen in the next chapter. In this section, we will study these types of martingales.

Recall that in Section 4.2 we showed that the stochastic process

$$Z_\lambda(t) = e^{\lambda W_{\mu,\sigma}(t)-\lambda\mu t-\frac{1}{2}\lambda^2\sigma^2 t} \quad (5.81)$$

is a martingale for any real number λ, where $\{W_{\mu,\sigma}(t)\}$ is Brownian motion with drift μ and volatility σ. Using this martingale, we derived other useful martingales in terms of Brownian motion. We now extend this result to Ito processes with the help of Ito's Lemma.

Assume that $\{X(t)\}$ be a solution to the following SDE

$$dX = \alpha(t, X)dt + \sigma(t, X)dW. \tag{5.82}$$

Let us begin by guessing what kind of exponential form involving $\{X(t)\}$ could be a martingale. (5.81) suggests that we consider

$$Z_b(t) = e^{\int_0^t b(s)dX(s) - \int_0^t b(s)\alpha(s, X(s))ds - \frac{1}{2}\int_0^t b^2(s)\sigma^2(s, X(s))ds}, \tag{5.83}$$

where $\{b(t)\}$ is a stochastic process. In other words, we may replace $\{W_{\mu,\sigma}(t)\}$ by $\{X(t)\}$, μ by $\alpha(t, X(t))$, σ by $\sigma(t, X(t))$, and λ by $b(t)$ in (5.81). Since $dX(t) = \alpha(t, X)ds + \sigma(t, X)dW$, we write (5.83) as

$$Z_b(t) = e^{\int_0^t b(s)\sigma(s, X(s))dW(s) - \frac{1}{2}\int_0^t b^2(s)\sigma^2(s, X(s))ds}. \tag{5.84}$$

We now show that under a certain condition $Z_b(t)$ is indeed a martingale. The condition which ensures that $Z_b(t)$ is a martingale is called Novikov's condition and is described as follows.

Consider a positive Ito process $\{Z(t)\}$. A necessary condition for the process $\{Z(t)\}$ to be a martingale is that its drift term vanishes, i.e.,

$$dZ(t) = \theta(t)Z(t)dW(t), \tag{5.85}$$

for some stochastic process $\{\theta(t)\}$. But not every process satisfying (5.85) is a martingale. Note that the volatilty of $\{Z(t)\}$ is $\sigma_Z(t, Z) = \theta(t)Z$. If $E\left\{\int_0^t \sigma_Z^2(s, Z(s))ds\right\}$ is finite for all t, i.e. (5.2) holds, then (5.6) implies that $\{Z(t)\}$ is a martingale. However, this condition is hard to verify for many positive martingales of interest. A commonly used sufficient condition for $\{Z(t)\}$ to be a martingale is

$$E\{e^{\frac{1}{2}\int_0^T [\theta(t)]^2 dt}\} < \infty. \tag{5.86}$$

The condition (5.86) is called Novikov's Condition. It can be shown (using advanced methods) that if (5.86) holds, then the stochastic process $\{Z(t)\}$ is a martingale. A weaker condition than Novikov's condition that also ensures that $\{Z(t)\}$ is a martingale is Kazamaki's Condition

$$E\{e^{\frac{1}{2}\int_0^t \theta(s)dW(s)}\} < \infty, \ 0 \le t \le T. \tag{5.87}$$

However, Kazamaki's condition is sometimes difficult to verify.

We are now ready to show that $\{Z_b(t)\}$ given in (5.84) is a martingale under a rather general condition.

Theorem 5.3 Let $\{X(t)\}$ be a solution of the SDE (5.82). For any stochastic process $\{b(t)\}$ for which the product $b(t)\sigma(t, X(t))$ satisfies Novikov's Condition, the stochastic process $\{Z_b(t)\}$ given in (5.84) is a martingale. $\{Z_b(t)\}$

is called an exponential martingale with respect to $\{X(t)\}$. In particular, if $\{b(t)\}$ and $\sigma(t, x)$ are bounded, $\{Z_b(t)\}$ is a martingale.

Proof: Define

$$Y(t) = -\frac{1}{2} \int_0^t b^2(s)\sigma^2(s, X(s))ds + \int_0^t b(s)\sigma(s, X(s))dW(s),$$

or

$$dY(t) = -\frac{1}{2}b^2(t)\sigma^2(t, X(t))dt + b(t)\sigma(t, X(t))dW(t).$$

So that, $Z_b(t) = e^{Y(t)}$. It follows from Ito's Lemma that

$$
\begin{aligned}
dZ_b(t) &= \frac{1}{2}b^2(t)\sigma^2(t, X(t))e^{Y(t)}dt + e^{Y(t)}dY(t) \\
&= \frac{1}{2}b^2(t)\sigma^2(t, X(t))Z_b(t)dt \\
&+ Z_b(t)\left[-\frac{1}{2}b^2(t)\sigma^2(t, X(t))dt + b(t)\sigma(t, X(t))dW(t)\right] \\
&= b(t)\sigma(t, X(t))Z_b(t)dW.
\end{aligned}
$$

Since $b(t)\sigma(t, X(t))$ satisfies Novikov's Condition, $\{Z_b(t)\}$ is a martingale.

$\square$

Theorem 5.3 is useful when we need to generate a martingale associated with an Ito process. As we have seen in Section 3.5 martingales play an important role in the change of probability measure. Theorem 5.3 provides a method to find a proper martingale such that the associated probability measure will meet certain required properties. This is the idea used in the proof of the Girsanov theorem in Chapter 6.

As in the Brownian motion case, we may also use Theorem 5.3 to construct other martingales. For any real number λ, let $b(t) = \lambda$. Thus, we obtain the martingale

$$Z_\lambda(t) = e^{\lambda X(t) - \lambda \int_0^t \alpha(s, X(s))ds - \frac{1}{2}\lambda^2 \int_0^t \sigma^2(s, X(s))ds}. \tag{5.88}$$

It follows upon double differentiation of $Z_\lambda(t)$ with respect to λ (and evaluation at $\lambda = 0$) that the two stochastic processes

$$X(t) - \int_0^t \alpha(s, X(s))ds \tag{5.89}$$

and

$$[X(t) - \int_0^t \alpha(s, X(s))ds]^2 - \int_0^t \sigma^2(s, X(s))ds \tag{5.90}$$

are martingales.

5.8 THE MARTINGALE REPRESENTATION THEOREM

In this section, we explore further the connection between martingales and Ito processes. Recall that an Ito process $\{X(t)\}$, which follows

$$dX = \alpha_X(t)dt + \sigma_X(t)dW,$$

where $\alpha_X(t)$ and $\sigma_X(t)$ are stochastic processes[10], is a stochastic process with respect to $\{\mathcal{B}_t\}$, the natural information structure generated by Brownian motion $\{W(t)\}$. Furthermore, $\{X(t)\}$ is a martingale only if its drift $\alpha_X(t) = 0$, as in the discussion of Novikov's Condition in the previous section. In other words, any Ito process that is also a martingale has the form

$$X(t) = X(0) + \int_0^t \sigma_X(s)dW(s). \tag{5.91}$$

The alternative question arises: If we are given a martingale $\{M(t)\}$ with respect to $\{\mathcal{B}_t\}$, under what conditions is $\{M(t)\}$ an Ito process in the form of (5.91)? This is not only of mathematical importance but also useful in the hedge of contingent claims as will be seen later in an example. It turns out that a sufficient condition is for $\{M(t)\}$ to be square integrable, i.e. for each t, $E\{M^2(t)\} < \infty$.

Theorem 5.4 The Martingale Representation Theorem

If a martingale $\{M(t)\}$ with respect to $\{\mathcal{B}_t\}$ is square integrable, then there is a unique square integrable stochastic process $\sigma_M(t)$ with respect to $\{\mathcal{B}_t\}$ such that

$$M(t) = M(0) + \int_0^t \sigma_M(s)dW(s). \tag{5.92}$$

The proof of this theorem is very complicated so we omit the details. In what follows, we outline the steps of the proof. First, consider the stochastic process

$$Z(t) = e^{\int_0^t h(s)dW(s) - \frac{1}{2}\int_0^t h^2(s)ds}, \tag{5.93}$$

where $h(s)$ is a deterministic function such that $\int_0^t h^2(s)ds$ is finite for all t. It is obvious that this is an exponential martingale as introduced in Section 5.6 with $b(s) = h(s)$. Thus, from the proof of Theorem 5.3

$$Z(t) = 1 + \int_0^t h(s)Z(s)dW(s), \tag{5.94}$$

i.e. $\sigma_Z(t) = h(t)Z(t)$. Next, extend (5.92) to linear combinations of exponential martingales of the above form. For

$$Y(t) = a_1 Z_1(t) + a_2 Z_2(t) + \cdots + a_k Z_k(t), \tag{5.95}$$

[10]This formula is slightly more general than that in Section 5.2 but it makes no difference in our discussion.

where $Z_i(t)$, $i = 1, 2, \cdots, k$, are in the form of (5.93), it follows from (5.94) that

$$Y(t) = Y(0) + \int_0^t \sum_{i=1}^{k} a_i h_i(s) Z_i(s) dW(s), \qquad (5.96)$$

where $h_i(t)$ is the corresponding deterministic function to $Z_i(t)$. Thus, $\sigma_Y(t) = \sum_{i=1}^{k} a_i h_i(t) Z_i(t)$. Finally, for an arbitrary martingale $\{M(t)\}$ satisfying the condition of Theorem 5.4, it can be shown using results in measure theory that there is a sequence $\{Y_n(t), n = 1, 2, \cdots\}$ of form (5.95) such that $M(t) = \lim_{n \to \infty} Y_n(t)$ in L^2. In this case, we show that $\{M(t)\}$ satisfies (5.92) with $\sigma_M(t) = \lim_{n \to \infty} \sigma_n(t)$, where $\sigma_n(t)$ is the volatility of $\{Y_n(t)\}$.

It is easy to see from the proof of Theorem 5.3 that the exponential martingale $Z_b(t)$ given in (5.84) satisfies

$$Z_b(t) = 1 + \int_0^t b(s)\sigma(s, X(s)) dW(s),$$

that is, $\sigma_Z(t) = b(t)\sigma(t, X(t))$.

Example 5.8 Consider the stochastic process

$$X(t) = e^{rt} \sin aW(t).$$

We will find the value of a, for which $\{X(t)\}$ is a martingale, and identify the volatility process.

Since $X(t)$ is a function of standard Brownian motion $\{W(t)\}$, Ito's Lemma applies. Therefore, we have

$$dX(t) = re^{rt} \sin[aW(t)]dt + e^{rt} d \sin[aW(t)]$$
$$= re^{rt} \sin[aW(t)]dt + e^{rt}\left\{-\frac{1}{2}a^2 \sin[aW(t)]dt + \cos[aW(t)]dW(t)\right\}$$
$$= \left(r - \frac{1}{2}a^2\right) e^{rt} \sin[aW(t)]dt + e^{rt} \cos[aW(t)]dW(t).$$

Thus, for $\{X(t)\}$ to be a martingale, one must have $r - \frac{1}{2}a^2 = 0$, i.e $a = \sqrt{2r}$. The volatility process is $\sigma_X(t) = e^{rt} \cos[\sqrt{2r}W(t)]$. $\qquad \square$

Theorem 5.4 also allows for a representation of random variables in terms of Ito integrals. Let X be a random variable on the probability space $(\Omega, \mathcal{B}_T, P)$ for some $T > 0$ with $E(X^2) < \infty$. Define $X(t) = E\{X|\mathcal{B}_t\}$. Then $\{X(t)\}$ is a martingale and from $E\{X^2(t)\} = E\{[E\{X|\mathcal{B}_t\}]^2\} \leq E(X^2)$ (by applying Jensen's inequality to $[E\{X|\mathcal{B}_t\}]^2$) it is square integrable. Therefore, Theorem 5.4 applies and one has, for $0 \leq t \leq T$,

$$E\{X|\mathcal{B}_t\} = E(X) + \int_0^t \sigma_X(s) dW(s). \qquad (5.97)$$

In particular,

$$X = E(X) + \int_0^T \sigma_X(s)dW(s). \tag{5.98}$$

The representation (5.97) has an important application in the identification of self-financing trading strategies as illustrated in the next example.

Example 5.9 Self-Financing Trading Strategy

Consider a risky asset whose price $\{S(t)\}$ follows

$$dS(t) = \alpha(t)S(t)dt + \sigma(t, S)dW(t), \tag{5.99}$$

where the deterministic function $\sigma(t, x) > 0$ for all $0 \le t \le T$ and $x > 0$.

Assume that $\{r(t)\}$ is the short rate process introduced in Section 5.3 following Example 5.6. Let $V(t) = e^{-\int_0^t r(u)du}S(t)$. Then $\{V(t)\}$ is the present value process of the stock price under the stochastic interest rates $\{r(t)\}$. We further assume that the market under consideration does not admit arbitrage. Thus, the Fundamental Theorem of Asset Pricing implies that there is a risk-neutral probability measure Q such that the present value process $\{V(t)\}$ is a martingale under this measure. Since

$$
\begin{aligned}
dV(t) &= -r(t)e^{-\int_0^t r(u)du}S(t)dt + e^{-\int_0^t r(u)du}dS(t) \\
&= [\alpha(t) - r(t)]V(t)dt + e^{-\int_0^t r(u)du}\sigma(t, S)dW(t),
\end{aligned}
$$

one must have $\alpha(t) = r(t)$ for all t, from the martingale property of $\{V(t)\}$. Thus, under the measure Q,

$$dS(t) = r(t)S(t)dt + \sigma(t, S)dW(t). \tag{5.100}$$

Consider now a contingent claim payable at time T. This contingent claim may be a European option maturing at T or a payoff contingent on the underlying Brownian motion process up to time T. Let $\Phi(T)$ be the payoff function of the contingent claim and assume that the second moment of $\Phi(T)$ is finite. The Fundamental Theorem of Asset Pricing shows that the fair time-t price of $\Phi(T)$ is given by

$$\Phi(t) = E_Q\left\{e^{-\int_t^T r(u)du}\Phi(T) \,\Big|\, \mathcal{B}_t\right\}, \quad 0 \le t \le T. \tag{5.101}$$

As seen in Section 5.4, in order for the writer of this claim to meet the obligation, one needs to construct a portfolio to replicate the price process $\{\Phi(t)\}$ with a self-financing trading strategy. In what follows, we apply the martingale representation theorem to identify such a strategy.

Introduce the stochastic process $M(t) = e^{-\int_0^t r(u)du} \Phi(t)$, $0 \le t \le T$. Then

$$M(t) = E_Q \left\{ e^{-\int_0^T r(u)du} \Phi(T) \,\Big|\, \mathcal{B}_t \right\}.$$

That is, $\{M(t)\}$ is a martingale since $X = e^{-\int_0^T r(u)du} \Phi(T)$ is independent of t. The martingale representation theorem thus yields

$$M(t) = M(0) + \int_0^t \sigma_M(s)dW(s), \tag{5.102}$$

or, in the differential form,

$$dM(t) = \sigma_M(t)dW(t).$$

On the other hand,

$$dM(t) = -r(t)e^{-\int_0^t r(u)du} \Phi(t)dt + e^{-\int_0^t r(u)du} d\Phi(t).$$

Define

$$\vartheta(t) = e^{\int_0^t r(u)du} \sigma_M(t)/\sigma(t, S(t)). \tag{5.103}$$

Then,

$$
\begin{aligned}
d\Phi(t) &= e^{\int_0^t r(u)du} dM(t) + r(t)\Phi(t)dt \\
&= e^{\int_0^t r(u)du} \sigma_M(t)dW(t) + r(t)\Phi(t)dt \\
&= \vartheta(t)\sigma(t, S(t))dW(t) + r(t)\Phi(t)dt \\
&= \vartheta(t)dS(t) - r(t)\vartheta(t)S(t)dt + r(t)\Phi(t)dt \\
&= \vartheta(t)dS(t) + r(t)[\Phi(t) - \vartheta(t)S(t)]dt.
\end{aligned}
$$

That is,

$$d\Phi(t) = \vartheta(t)dS(t) + r(t)[\Phi(t) - \vartheta(t)S(t)]dt. \tag{5.104}$$

The interpretation of (5.104) is very similar to that of (5.49). Let $\vartheta(t)$ given in (5.103) represent the number of shares of the risky asset retained at time t. If a portfolio consisting of the risky asset and the money market account replicates the price process $\{\Phi(t)\}$, its time-t value is equal to $\Phi(t)$. Since $\vartheta(t)$ is the number of shares at time t, the value of the risky asset is $\vartheta(t)S(t)$ and therefore the value in the money market account that earns interest at rate $r(t)$ is $\Phi(t) - \vartheta(t)S(t)$. Thus, the first term on the right-hand side of (5.104) is the capital gain from the risky asset over the period $[t, t + dt)$ and the second term is the gain from the money market account. Since the left-hand side of (5.104) represents the increase in the price of the contingent claim, the relation (5.104) implies that the total gains from the market are fully reinvested. In other words, the self-financing trading strategy which retains $\vartheta(t)$ shares of the risky asset at time t replicates the price of the contingent claim throughout the entire period $[0, T]$. $\square$

CHAPTER 6

STOCHASTIC CALCULUS: ADVANCED TOPICS

In this chapter, we discuss topics that have important financial applications but are more mathematically challenging than those in Chapter 5. We begin with the Feynman-Kac formula which provides a probabilistic solution to certain types of partial differential equations (PDEs). As an application, the Black-Scholes PDE derived in Section 5.4 is solved using the Feynman-Kac formula. The Girsanov theorem is the subject of Section 6.3. The Girsanov theorem has many applications, and often plays a central role in evaluating contingent claims and derivative securities in particular. The risk-neutral valuation of contingent claims requires a probability measure equivalent to the physical probability measure, such that the present value process of underlying securities is a martingale, as seen in Section 3.5 in a discrete-time framework. When continuous-time models are considered, the Girsanov theorem is an appropriate tool to find such a probability measure. Section 6.4 deals with complex barrier hitting probabilities. We provide a detailed derivation in order to illustrate the use of the Girsanov theorem and the further use of the reflection principle. In the final section, we present results of stochastic calculus in a two-dimensional setting. These results are counterparts to those in the one-dimensional case but contain additional terms that reflect correlations between two stochastic processes. The proofs of the results are similar to those in the one-dimensinal case and are thus omitted. The extension to higher dimensions from two dimensions is rather trivial, which is why it is not included in this chapter.

6.1   THE FEYNMAN-KAC FORMULA

Partial differential equations play an important role in pricing contingent claims. As we have seen in Section 5.4, the price of a European option under the Black-Scholes setting is expressed as the solution to a PDE using the self-financing argument. This is also the case when the underlying security on which a contingent claim is written follows a more general Ito process and the interest rate is stochastic. The Feynman-Kac formula that gives a probabilistic representation of the solution to certain types of PDEs may thus provide a closed form expression for the price of the contingent claim in these cases.

Theorem 6.1 The Feynman-Kac Formula

Let $u(t, x)$ be the solution to the following partial differential equation (PDE)

$$\frac{\partial}{\partial t}u(t, x) + \frac{1}{2}\frac{\partial^2}{\partial x^2}u(t, x) + \gamma(x)u(t, x) = 0, \ 0 \le t \le T. \qquad (6.1)$$

with terminal condition $u(T, x) = \psi(x)$, where $\gamma(x)$ is a continuous function bounded from above, and $|u(t, x)| \le Le^{|x|^k}$, for some $L > 0$ and $k < 2$. Then

$$u(t, x) = E\left\{ e^{\int_t^T \gamma(W(s))ds}\psi(W(T)) \ \middle| \ W(t) = x\right\}, \qquad (6.2)$$

where $\{W(t)\}$ is standard Brownian motion.

Proof: Denote

$$X_1(t) = e^{\int_0^t \gamma(W(s))ds}, \ \text{ and } \ X_2(t) = u(t, W(t)).$$

Then

$$dX_1 = \gamma(W)X_1 dt$$

and

$$dX_2 = \left[\frac{\partial}{\partial t}u(t, W) + \frac{1}{2}\frac{\partial^2}{\partial x^2}u(t, W)\right]dt + \frac{\partial}{\partial x}u(t, W)dW.$$

The first equation is obtained by ordinary differentiation and the second equation is obtained by Ito's Lemma.

Integration by parts yields

$$\begin{aligned} d[X_1 X_2] &= X_2 dX_1 + X_1 dX_2 \\ &= e^{\int_0^t \gamma(W(s))ds} \frac{\partial}{\partial x} u(t, W) dW. \end{aligned}$$

Thus the process $\{X_1(t)X_2(t)\}$ has zero drift. It can be shown that under the condition of the theorem, $\{X_1(t)X_2(t)\}$ is a martingale. Thus,

$$E\left\{ e^{\int_0^T \gamma(W(s))ds} u(T, W(T)) \;\middle|\; W(t) = x \right\} = e^{\int_0^t \gamma(W(s))ds} u(t, x).$$

Dividing both sides by $e^{\int_0^t \gamma(W(s))ds}$ yields the result, since $u(T, W(T)) = \psi(W(T))$. $\qquad\square$

We remark:

i) Although the probabilistic representation (6.2) for the solution to Equation (6.1) requires that the conditions on $\gamma(x)$ and $u(t, x)$ be met, one should not worry about them at first. A typical procedure is to use formula (6.2) to obtain a candidate for the solution to (6.1) and to then check whether or not it satisfies these conditions;

ii) The Feynman-Kac formula in most probability literature is derived for much more general nonhomogeneous PDEs, defined on a multi-dimensional domain. See Karatzas and Shreve (1991), p. 366 for example.

6.2 THE BLACK-SCHOLES PARTIAL DIFFERENTIAL EQUATION

In this section, we apply the Feynman-Kac formula to solve the Black-Scholes PDE derived in Section 5.4. As in Section 5.4, if one assumes the Black-Scholes economy in which there is a money market account that earns interest at constant rate r and a risky asset $\{S(t)\}$ that follows a geometric Brownian motion process with volatility σ, then the self-financing (also called the delta hedging argument) results in the following Black-Scholes PDE

$$\frac{\partial}{\partial t}\phi(t, S) + \frac{1}{2}\sigma^2 S^2 \frac{\partial^2}{\partial S^2}\phi(t, S) + rS\frac{\partial}{\partial S}\phi(t, S) - r\phi(t, S) = 0 \qquad (6.3)$$

with terminal condition

$$\phi(T, S) = \Phi(S), \qquad (6.4)$$

for any $0 \le t \le T$ and $S \ge 0$. Here the solution $\phi(t, S)$ is the time-t price of the option which pays $\Phi(S(T))$ at maturity time T, when the value of the underlying risky asset at time t is S.

In Section 5.4, we gave the solution to (6.3) without a proof. We now demonstrate how to use the Feynman-Kac Formula given in Section 6.1 to solve Equation (6.3). It is obvious that we can not apply this formula directly. In the following two steps, we use analytical tools to convert the PDE (6.3) into a PDE to which the Feynman-Kac Formula does apply.

1. Eliminating S and S^2 from the coefficients of the partial derivatives in (6.3).

Noting that the PDE is similar to an Euler-type ODE, let $S = e^{\sigma x}$ and $v(t,x) = \phi(t, e^{\sigma x})$. Then,

$$\frac{\partial}{\partial x}v(t,x) = \sigma e^{\sigma x}\frac{\partial}{\partial x}\phi(t, e^{\sigma x}) = \sigma S\frac{\partial}{\partial S}\phi(t, S),$$

$$\frac{\partial^2}{\partial x^2}v(t,x) = \sigma^2 S^2\frac{\partial^2}{\partial S^2}\phi(t, S) + \sigma^2 S\frac{\partial}{\partial S}\phi(t, S).$$

Substitution of partial derivatives of $v(t,x)$ into (6.3) shows that $v(t,x)$ is the solution to the PDE

$$\frac{\partial}{\partial t}v(t,x) + \frac{1}{2}\frac{\partial^2}{\partial x^2}v(t,x) + \frac{1}{\sigma}\left(r - \frac{1}{2}\sigma^2\right)\frac{\partial}{\partial x}v(t,x) - rv(t,x) = 0, \quad (6.5)$$

and it follows from the terminal condition (6.4) that $v(T,x) = \Phi(e^{\sigma x})$.

2. Eliminating the first-order term $\frac{\partial}{\partial x}v(t,x)$ from (6.5).

Since (6.5) is similar to a second-order linear ODE, the technique used in solving the second-order linear ODE may apply. Let $v(t,x) = e^{\kappa x}u(t,x)$, where κ will be determined later. Then,

$$\frac{\partial}{\partial x}v(t,x) = e^{\kappa x}\frac{\partial}{\partial x}u(t,x) + \kappa e^{\kappa x}u(t,x),$$

$$\frac{\partial^2}{\partial x^2}v(t,x) = e^{\kappa x}\frac{\partial^2}{\partial x^2}u(t,x) + 2\kappa e^{\kappa x}\frac{\partial}{\partial x}u(t,x) + \kappa^2 e^{\kappa x}u(t,x).$$

Thus

$$\frac{\partial}{\partial t}u(t,x) + \frac{1}{2}\frac{\partial^2}{\partial x^2}u(t,x) + \left[\kappa + \frac{1}{\sigma}\left(r - \frac{1}{2}\sigma^2\right)\right]\frac{\partial}{\partial x}u(t,x)$$
$$+ \left[\frac{1}{2}\kappa^2 + \frac{\kappa}{\sigma}\left(r - \frac{1}{2}\sigma^2\right) - r\right]u(t,x) = 0. \quad (6.6)$$

Let $\kappa = -\frac{1}{\sigma}(r - \frac{1}{2}\sigma^2)$. Then, $u(t,x)$ is the solution to the PDE

$$\frac{\partial}{\partial t}u(t,x) + \frac{1}{2}\frac{\partial^2}{\partial x^2}u(t,x) - \left[\frac{1}{2\sigma^2}\left(r - \frac{1}{2}\sigma^2\right)^2 + r\right]u(t,x) = 0, \quad (6.7)$$

with $u(T,x) = e^{\frac{1}{\sigma}(r-\frac{1}{2}\sigma^2)x}\Phi(e^{\sigma x})$.

We are ready to apply the Feynman-Kac Formula. Comparison of (6.1) with
(6.7) yields $\gamma(x) = -[\frac{1}{2\sigma^2}(r - \frac{1}{2}\sigma^2)^2 + r]$ and $\psi(x) = e^{\frac{1}{\sigma}(r-\frac{1}{2}\sigma^2)x}\Phi(e^{\sigma x})$.
Hence, the Feynman-Kac formula (6.2) implies

$$u(t,x) = e^{-[\frac{1}{2\sigma^2}(r-\frac{1}{2}\sigma^2)^2+r](T-t)}E\left\{ e^{\frac{1}{\sigma}(r-\frac{1}{2}\sigma^2)W(T)}\Phi(e^{\sigma W(T)}) \Big| W(t) = x\right\}.$$
$$(6.8)$$

Note that $W(T)$, given that $W(t) = x = \frac{1}{\sigma}\log S$, is a normal random
variable with mean x and variance $T - t$. We obtain

$$
\begin{aligned}
\phi(t,S) &= e^{-[\frac{1}{2\sigma^2}(r-\frac{1}{2}\sigma^2)^2+r](T-t)-\frac{1}{\sigma}(r-\frac{1}{2}\sigma^2)x}\\
&\times E\left\{ e^{\frac{1}{\sigma}(r-\frac{1}{2}\sigma^2)W(T)}\Phi(e^{\sigma W(T)}) \Big| W(t) = x\right\}\\
&= \frac{1}{\sqrt{2\pi(T-t)}}e^{-[\frac{1}{2\sigma^2}(r-\frac{1}{2}\sigma^2)^2+r](T-t)-\frac{1}{\sigma}(r-\frac{1}{2}\sigma^2)x}\\
&\times \int_{-\infty}^{\infty} e^{\frac{1}{\sigma}(r-\frac{1}{2}\sigma^2)y}\Phi(e^{\sigma y})e^{-\frac{1}{2(T-t)}(y-x)^2}dy\\
&= \frac{e^{-r(T-t)}}{\sqrt{2\pi(T-t)}}\int_{-\infty}^{\infty}\Phi(Se^{\sigma y})e^{-\frac{1}{2(T-t)}[y-\frac{1}{\sigma}(r-\frac{1}{2}\sigma^2)(T-t)]^2}dy\\
&= \frac{e^{-r(T-t)}}{\sqrt{2\pi(T-t)}}\int_{-\infty}^{\infty}\Phi(Se^{(r-\frac{1}{2}\sigma^2)(T-t)+\sigma y})e^{-\frac{1}{2(T-t)}y^2}dy.
\end{aligned}
$$

Thus we have derived the formula (5.55) in Section 5.4..

6.3 THE GIRSANOV THEOREM

In this section, we discuss the Girsanov Theorem, a fundamental result in
stochastic calculus. Roughly speaking, the Girsanov Theorem states that for
a given Ito process, i.e., a solution to some SDE, we may adjust the probability
of each path of the process so that the Ito process under the new probabilities
has a specified drift. This theorem has many applications in option pricing
theory. Pricing an option or a contingent claim often requires finding a probability measure under which the underlying asset price process has the same
stochastic return as that of the money market account or a numeraire process
of our choice (a long-term zero-coupon bond for example). The Girsanov
Theorem provides a useful tool to find such a probability measure.

The Girsanov Theorem Version A

Let $\{W(t), 0 \leq t \leq T\}$, be standard Brownian motion on the usual filtered
probability space $(\Omega, \{\mathcal{B}_t\}, P)$, where $\{\mathcal{B}_t\}$ is the natural information structure with respect to Brownian motion $\{W(t)\}$. Also, let $\{b(t)\}$ be a stochastic

process with respect to $\{\mathcal{B}_t\}$, satisfying Novikov's condition (5.86). Define a real-valued function $Q(\cdot)$ on all the events in $\mathcal{B}_T$ as follows. For any $F \in \mathcal{B}_T$,

$$Q(F) = E\{\mathbf{I}_F Z_b(T)\} = \int_F Z_b(T)dP, \tag{6.9}$$

where $Z_b(T) = e^{\int_0^T b(t)dW(t) - \frac{1}{2}\int_0^T b^2(t)dt}$. Then,

1. Q is a probability measure on $(\Omega, \mathcal{B}_T)$ that is equivalent to P, i.e., for any event F the probability of F is non-zero under P if and only if the probability of F is non-zero under Q. In measure-theory terminology, $Z_b(T)$ is referred to as the Radon-Nikodym derivative $\frac{dQ}{dP}$ of the probability measure Q with respect to the probability measure P; item the stochastic process

$$\tilde{W}(t) = -\int_0^t b(s)ds + W(t), \; 0 \leq t \leq T, \tag{6.10}$$

is standard Brownian motion under the probability measure Q.

We now give some intuitive explanations of the theorem before proceeding with its proof. The right-hand side of (6.10), $-\int_0^t b(s)ds + W(t)$, represents a stochastic process with a predetermined drift (process) $-\int_0^t b(s)ds$ under the original probability measure P since $\{W(t)\}$ has zero drift under P. In order to make the drift disappear, we adjust the probability of each path according to (6.9), where F may be viewed as a path up to time T. More precisely, the Radon-Nikodym derivative $Z_b(T)$ is the adjustment factor and the new probability of the path, $Q(F)$, is equal to the product of $Z_b(T)(F)$ and the old probability $P(F)$.

Proof: Define the stochastic process

$$Z_b(t) = e^{\int_0^t b(s)dW(s) - \frac{1}{2}\int_0^t b^2(s)ds}. \tag{6.11}$$

Since $\{b(t)\}$ satisfies Novikov's Condition, it follows from Theorem 5.3 that $\{Z_b(t)\}$ is an exponential martingale under P (since $\alpha = 0$ and $\sigma = 1$ in this case). Thus, the martingale property yields

$$Q(\Omega) = E(Z_b(T)) = Z_b(0) = 1.$$

It is easy to see from the definition (6.9) that Q is nonnegative and additive on $\mathcal{F}_T$. Hence Q is a probability measure. Since $Z_b(T)$ is always finite and positive, the two probability measures Q and P are equivalent.

Next, we show that $\tilde{W}(t)$ given in (6.10) is a standard Brownian motion process under the probability measure Q. Recall (the martingale property of Brownain motion, Section 4.5) that $\tilde{W}(t)$ is a standard Brownian motion process if and only if for any real number λ,

$$Z_\lambda(t) = e^{\lambda \tilde{W}(t) - \frac{1}{2}\lambda^2 t}$$

is a martingale. Denote by E_Q the expectation under Q. We need to show that for any $s > t$,

$$E_Q\{Z_\lambda(s) \mid \mathcal{B}_t\} = Z_\lambda(t). \tag{6.12}$$

Given any $F \in \mathcal{B}_t$, one has

$$
\begin{aligned}
\int_F E_Q\{Z_\lambda(s) \mid \mathcal{B}_t\} dQ &= \int_F E_Q\{Z_\lambda(s) \mid \mathcal{B}_t\} Z_b(T) dP \\
&= \int_F E\{E_Q\{Z_\lambda(s) \mid \mathcal{B}_t\} Z_b(T) \mid \mathcal{B}_t\} dP \\
&= \int_F E_Q\{Z_\lambda(s) \mid \mathcal{B}_t\} Z_b(t) dP.
\end{aligned}
$$

On the other hand,

$$
\begin{aligned}
\int_F E_Q\{Z_\lambda(s) \mid \mathcal{B}_t\} dQ &= \int_F Z_\lambda(s) dQ = \int_F Z_\lambda(s) Z_b(T) dP \\
&= \int_F E\{Z_\lambda(s) Z_b(T) \mid \mathcal{B}_t\} dP \\
&= \int_F E\{E\{Z_\lambda(s) Z_b(T) \mid \mathcal{B}_s\} \mid \mathcal{B}_t\} dP \\
&= \int_F E\{Z_\lambda(s) Z_b(s) \mid \mathcal{B}_t\} dP.
\end{aligned}
$$

Since both $E_Q\{Z_\lambda(s) \mid \mathcal{B}_t\} Z_b(t)$ and $E\{Z_\lambda(s) Z_b(s) \mid \mathcal{B}_t\}$ are random variables with respect to $\mathcal{B}_t$, it follows from the uniqueness of the conditional expectation that

$$E_Q\{Z_\lambda(s) \mid \mathcal{B}_t\} Z_b(t) = E\{Z_\lambda(s) Z_b(s) \mid \mathcal{B}_t\}. \tag{6.13}$$

We now show that $\{Z_\lambda(t) Z_b(t)\}$ is a martingale under P. Since

$$Z_\lambda(t) = e^{\lambda W(t) - \lambda \int_0^t b(s) ds - \frac{1}{2}\lambda^2 t}$$

and

$$Z_b(t) = e^{\int_0^t b(s) dW(s) - \frac{1}{2} \int_0^t b^2(s) ds},$$

$$
\begin{aligned}
Z_\lambda(t) Z_b(t) &= e^{\int_0^t [\lambda + b(s)] dW(s) - \frac{1}{2}[\lambda^2 t + 2\lambda \int_0^t b(s) ds + \int_0^t b^2(s) ds]} \\
&= e^{\int_0^t [\lambda + b(s)] dW(s) - \frac{1}{2} \int_0^t [\lambda + b(s)]^2 ds},
\end{aligned}
$$

i.e. $Z_\lambda(t)Z_b(t)$ is an exponential martingale with $b_*(t) = \lambda + b(t)$ under P. Thus, $Z_\lambda(t)Z_b(t)$ is a martingale and

$$E\Big\{Z_\lambda(s)Z_b(s) \mid \mathcal{B}_t\Big\} = Z_\lambda(t)Z_b(t).$$

It follows from (6.13) that

$$E_Q\Big\{Z_\lambda(s) \mid \mathcal{B}_t\Big\}Z_b(t) = E\Big\{Z_\lambda(s)Z_b(s) \mid \mathcal{B}_t\Big\} = Z_\lambda(t)Z_b(t).$$

Since $Z_b(t)$ is strictly positive, we have

$$E_Q\Big\{Z_\lambda(s) \mid \mathcal{B}_t\Big\} = Z_\lambda(t).$$

That is, $\{Z_\lambda(t)\}$ is a martingale under the probability measure Q. Therefore, the stochastic process $\{\tilde{W}(t)\}$ is standard Brownian motion under Q and the proof is complete. $\qquad\square$

Some observations can be made from the proof of the Girsanov Theorem. First, for each t, the random variable $Z_b(t)$ is the Radon-Nikodym derivative of the measure P with respect to the measure Q on the probability space $(\Omega, \mathcal{B}_t)$. In other words, $Z_b(t)$ is the adjustment factor for paths up to time t. Second, it follows from (6.10) that

$$d\tilde{W}(t) = -b(t)dt + dW(t). \tag{6.14}$$

Since

$$Z_b^{-1}(t) = e^{-\int_0^t b(s)dW(s)+\frac{1}{2}\int_0^t b^2(s)ds},$$

replacing $dW(s)$ by $b(s)ds + d\tilde{W}(s)$ yields that the inverse process $Z_b^{-1}(t)$ can be written as

$$Z_b^{-1}(t) = e^{-\int_0^t b(s)d\tilde{W}(s)-\frac{1}{2}\int_0^t b^2(s)ds}. \tag{6.15}$$

Since $\tilde{W}(t)$ is standard Brownian motion under the probability measure Q, $\{Z_b^{-1}(t)\}$ is an exponential martingale under the probability measure Q. That is, $\{Z_b(t)\}$ is a P-martingale and $\{Z_b^{-1}(t)\}$ is a Q-martingale. This property is often used when we need to change from the probability measure Q back to the probability measure P and vice versa.

We now examine the SDE

$$dX = \alpha(t, X)dt + \sigma(t, X)dW, \ 0 \le t \le T, \tag{6.16}$$

under the new probability measure Q.

It follows from (6.14) that

$$
\begin{aligned}
dX &= \alpha(t, X)dt + \sigma(t, X)[b(t)dt + d\tilde{W}(t)] \\
&= [\alpha(t, X) + \sigma(t, X)b(t)]dt + \sigma(t, X)d\tilde{W}. \qquad (6.17)
\end{aligned}
$$

In other words, under the probability measure Q, the SDE (6.16) has a new drift $\alpha(t, X) + \sigma(t, X)b(t)$, but maintains the same volatility $\sigma(t, X)$. This leads to the second version of the Girsanov Theorem.

The Girsanov Theorem Version B

Let $X(t)$ be a solution to the SDE (6.16), where $\sigma(t, x)$ is a positive function. Also let $\beta(t, x)$ be a continuous function such that

$$
\frac{\beta(t, x) - \alpha(t, x)}{\sigma(t, x)} \qquad (6.18)
$$

is bounded or satisfies Novikov's Condition.

Then, there exists a probability measure Q under which $X(t)$ is a solution to the following SDE

$$
dX = \beta(t, X)dt + \sigma(t, X)d\tilde{W}, \qquad (6.19)
$$

where $\tilde{W}(t)$ is standard Brownian motion under Q. Moreover, the Radon-Nikodym derivative of Q with respect to P is

$$
\frac{dQ}{dP} = e^{\int_0^T b(t)dW(t) - \frac{1}{2}\int_0^T b^2(t)dt}, \qquad (6.20)
$$

where

$$
b(t) = \frac{\beta(t, X(t)) - \alpha(t, X(t))}{\sigma(t, X(t))}. \qquad (6.21)
$$

The Version B of the Theorem is a very important result for continuous-time finance models. As we mentioned at the beginning of this section, pricing a contingent claim often requires us to find a probability measure for which the underlying risky asset has a specified drift. Version B provides a method to find such a probability measure.

6.4   THE FORWARD RISK ADJUSTED MEASURE AND BOND OPTION PRICING

In this section, we apply the Girsanov Theorem to derive the so-called forward risk adjusted measure and a closed-form expression for the price of options on coupon bonds. The work presented in this section is due to Jamshidian (1989, 1991).

Consider again the one-factor Gaussian forward rate model (5.41) introduced in Section 5.4:

$$df(t, s) = \alpha(t, s)dt + \sigma(t, s)\, dW(t),\ 0 \le t \le s, \qquad (6.22)$$

where $\sigma(t, s)$ is deterministic and

$$\alpha(t, s) = \sigma(t, s) \int_t^s \sigma(t, y)dy. \qquad (6.23)$$

under the risk-neutral measure Q.

The short rate process $\{r(t)\}$ then satisfies

$$dr(t) = \alpha(t)dt + \sigma(t, t)\, dW(t), \qquad (6.24)$$

where $\alpha(t)$ is given in (5.46). For any $\mathcal{B}_s$-contingent claim $C(s)$, payable at time s, its price $C(t)$, $t \le s$, can be obtained via the formula (5.33). However, the implementation of the formula (5.33) is some time difficult because the joint distibution of $e^{-\int_t^s r(u)du}$ and $C(s)$ under the risk-neutral measure Q needs to be identified. That is often the case when the payoff function of a contingent claim is complex. This problem may be solved in some situations by changing the risk neutral measure to the forward risk adjusted measure introduced in Jamshidian (1989). Rewrite (6.22) as

$$df(t, s) = \sigma(t, s) \left[\int_t^s \sigma(t, y)dydt + dW(t) \right],\ 0 \le t \le s, \qquad (6.25)$$

using (6.23). This suggests that if we let, for $0 \le t \le s$,

$$W_s(t) = \int_0^t \int_u^s \sigma(u, y)dydu + W(t), \qquad (6.26)$$

or

$$b(t, s) = - \int_t^s \sigma(t, y)dy, \qquad (6.27)$$

then, the Girsanov Theorem with $b(t) = b(t, s)$, $\le t \le s$, implies that for the period $[0,\ s]$, there is a probability measure Q_s such that $W_s(t)$, $0 \le t \le s$ is standard Brownian motion under Q_s. The measure Q_s is referred to as the forward risk adjusted measure. The SDE (6.22) or (6.25) can then be rewritten as

$$df(t, s) = \sigma(t, s)\, dW_s(t),\ 0 \le t \le s. \qquad (6.28)$$

In this case, the forward rate process $f(t, s)$, $0 \le t \le s$, is a martingale under the forward risk adjusted measure Q_s. Furthermore, A useful martingale property related to the bond price $P(t, s)$ can be derived. It is easy to see that the bond price $P(t, s)$ satisfies

$$
\begin{aligned}
dP(t, s) &= r(t)P(t, s)dt + b(t, s)P(t, s)dW(t) & (6.29)\\
&= \left\{ r(t) + [b(t, s)]^2 \right\} P(t, s)dt + b(t, s)P(t, s)dW_s(t) & (6.30)
\end{aligned}
$$

As shown in Example 5.9, the Martingale Representation Theorem implies that the price process $C(t)$, $0 \le t \le s$, obtained by

$$C(t) = E_Q \left\{ e^{-\int_t^s r(u)du} C(s) \,\bigg|\, \mathcal{B}_t \right\}$$

satisfies

$$dC(t) = r(t)C(t)dt - \sigma_C(t)C(t)dW(t), \tag{6.31}$$

for some volatility process $\sigma_C(t)$, $0 \le t \le s$. It follows from (5.76), (6.29) and (6.31) that

$$
\begin{aligned}
d\frac{C(t)}{P(t,s)} &= \frac{1}{P(t,s)}dC(t) - \frac{C(t)}{[P(t,s)]^2}dP(t,s) \\
&+ \left[\frac{[b(t,s)P(t,s)]^2 C(t)}{[P(t,s)]^3} + \frac{[\sigma_C(t)C(t)][b(t,s)P(t,s)]}{[P(t,s)]^2} \right] dt \\
&= \frac{C(t)}{P(t,s)} \{ b(t,s)[b(t,s) + \sigma_C(t)]dt \\
&- [b(t,s) + \sigma_C(t)]dW(t) \} \\
&= -\frac{C(t)}{P(t,s)}[b(t,s) + \sigma_C(t)]dW_s(t). \tag{6.32}
\end{aligned}
$$

Thus, $C(t)/P(t,s)$, $0 \le t \le s$, is a martingale under the forward risk adjusted measure Q_s. This martingale property enables us to write

$$\frac{C(t)}{P(t,s)} = E_s \left\{ \frac{C(s)}{P(s,s)} \,\bigg|\, \mathcal{B}_t \right\} = E_s \{C(s)|\mathcal{B}_t\},$$

where $E_s(\cdot)$ is the expectation with respect to the forward risk adjusted measure Q_s. Therefore,

$$C(t) = P(t,s)E_s\{C(s)\,|\mathcal{B}_t\}. \tag{6.33}$$

The formula (6.33) has an obvious advantage: the discount process $e^{-\int_t^s r(u)du}$ is separated from the contingent claim payoff under the forward risk adjusted measure Q_s. Hence, we do not need to know the joint distribution of $e^{-\int_t^s r(u)du}$ and $C(s)$ in order to calculate the value of $C(t)$. Instead, we only need to know the marginal distribution of $C(s)$ under the measure Q_s. The formula (6.33) is useful for pension valuation for which one often needs to valuate a fixed income portfolio. A common practice is to calculate the expected cash flow from the portfolio and then to discount the expected cash flow using a yield curve. The formula (6.33) suggests that to do so one must use forward risk adjusted measures, not the physical probability measure nor the risk-neutral probability measure, to calculate the expectations.

In the following, we use (6.33) to derive closed form expressions for the prices of bond options. First consider European call and put options on the zero-coupon bond $P(t, T)$ with strike price K and expiration time s, $t \leq s \leq T$. The payoff of the option is

$$\max\{P(s, T) - K, \, 0\}. \qquad (6.34)$$

Suppose that the forward rates follow the one-factor Gaussian model (6.22). As shown above, the process $P(t, T)/P(t, s)$, $0 \leq t \leq s$, is a martingale under the forward risk adjusted measure Q_s and satisfies

$$
\begin{aligned}
d\frac{P(t, T)}{P(t, s)} &= \frac{P(t, T)}{P(t, s)} \left[\int_t^s \sigma(t, y) dy - \int_t^T \sigma(t, y) dy \right] dW_s(t) \\
&= -\frac{P(t, T)}{P(t, s)} \left[\int_s^T \sigma(t, y) dy \right] dW_s(t). \qquad (6.35)
\end{aligned}
$$

Hence, $P(s, T) = P(s, T)/P(s, s)$ is lognormal with log-variance

$$[v(t, s, T)]^2 = \int_t^s \left[\int_s^T \sigma(u, y) dy \right]^2 du, \qquad (6.36)$$

and log-mean

$$\ln(P(t, T)/P(t, s)) - \frac{1}{2}[v(t, s, T)]^2.$$

To calculate $E_s\{\max\{P(s, T) - K, \, 0\}\}$, we apply (5.63) with

$$\bar{\mu} = \ln(P(t, T)/P(t, s)) - \frac{1}{2}[v(t, s, T)]^2$$

and

$$\bar{\sigma}^2 = [v(t, s, T)]^2.$$

Since

$$e^{\bar{\mu} + \frac{1}{2}\bar{\sigma}^2} = \frac{P(t, T)}{P(t, s)},$$

and

$$\frac{\bar{\mu} - \ln K + \bar{\sigma}^2}{\bar{\sigma}} = \frac{\ln(P(t, T)/P(t, s)K) + \frac{1}{2}v(t, s, T)^2}{v(t, s, T)},$$

it follows from (6.33) and (5.63) that the price of the call option is

$$
\begin{aligned}
\phi_c(t) &= P(t, s) E_s\{\max\{P(s, T) - K, \, 0\}\} \\
&= P(t, s) \left\{ \frac{P(t, T)}{P(t, s)} N(d_1(t, s, T)) - KN(d_2(t, s, T)) \right\} \\
&= P(t, T)N(d_1(t, s, T)) - KP(t, s)N(d_2(t, s, T)),
\end{aligned}
$$

$$(6.37)$$

where

$$d_1(t, s, T) = \frac{\ln(P(t, T)/P(t, s)K) + \frac{1}{2}v(t, s, T)^2}{v(t, s, T)} \qquad (6.38)$$

and

$$d_2(t, s, T) = d_1(t, s, T) - v(t, s, T). \qquad (6.39)$$

The corresponding put price may be obtained by the put-call parity and it is given by

$$
\begin{aligned}
\phi_p(t) &= P(t, s)E_s\{\max\{K - P(s, T), 0\}\} \\
&= -P(t, T)N(-d_1(t, s, T)) + KP(t, s)N(-d_2(t, s, T)),
\end{aligned}
$$
$$\qquad (6.40)$$

where $d_1(t, s, T)$ and $d_2(t, s, T)$ are given in (6.38) and (6.39).

Consider now European call and put options on a coupon bond. In order to derive closed form expressions for the options, some extra conditions are to imposed. We assume that the term structure of volatilities for the forward rates is of the form

$$\sigma(t, s) = \sigma(t, t)e^{-\int_t^s \kappa(u)du}, \qquad (6.41)$$

where $\kappa(t)$ is a non-negative deterministic function. The function $\kappa(t)$ may be interpreted as the intensity of the forward rate volatility. Note that the extended Vasicek model (Section 5.4) is a special case of (6.41).

Differentiating (6.41) with respect to s, we have

$$\frac{\partial \sigma(t, s)}{\partial s} = -\kappa(t)\sigma(t, s).$$

We can thus write the SDE (6.24) as

$$dr(t) = \kappa(t)[\theta(t) - r(t)]dt + \sigma(t, t) \, dW(t), \qquad (6.42)$$

similar to the derivation in the end of Section 5.4. Solving (6.42) and then using (5.40), we can write the bond price $P(t, s)$ as

$$P(t, s) = e^{A(t,s) - B(t,s)r(t)}, \qquad (6.43)$$

where $A(t, s)$ and $B(t, s)$ are deterministic and moreover $B(t, s)$ is positive. A short rate model under which the bond prices $P(t, s)$, for all $t \leq s$, can be expressed in the form of (6.43) is called an affine model. Note that an affine model can be Gaussian or non-Gaussian and the model we discuss here is a very special case. For affine models, see Duffie and Kan (1996) and Cairns (2004).

With the help of (6.43), we can now express the bond price $P(t, s)$ as

$$P(t, s) = h(r(t), t, s), \qquad (6.44)$$

where

$$h(r,t,s) = e^{A(t,s)-B(t,s)r} \tag{6.45}$$

is a deterministic function and is decreasing in r.

Let $F(t,T)$ is the time-t price of a coupon bond with maturity time T as described in Section 3.7. The coupon bond has a par value F and coupon C_i payable at time $s_i, i = 1, \cdots, j$, where $0 < s_1 < s_2 < \cdots < s_j = T$. Thus, $F(t,T)$ may be expressed in terms of a series of zero coupon bonds:

$$F(t,T) = \sum_{s_i > t} C_i P(t,s_i) + F\,P(t,T). \tag{6.46}$$

Similar to (6.34), we consider a European call option on the coupon bond that pays

$$\max\{F(s,T) - K, 0\} \tag{6.47}$$

at time s. To derive a closed form expression for the price of the option, first solve the equation

$$\sum_{s_i \geq s} C_i h(r,s,s_i) + F\,h(r,s,T) = K$$

for r. Since each function $h(r,s,s_i)$ is decreasing, the equation has a unique solution. Let r^* be the solution and let let $K_i = h(r^*, s, s_i)$, $s_i \geq s$. Again, the decreasing property of function $h(r,s,s_i)$ in r allows to write

$$\max\{F(s,T) - K, 0\}$$
$$= \sum_{s_i \geq s} C_i \max\{P(s,s_i) - K_i, 0\} + F\,\max\{P(s,T) - K_j, 0\}.$$

In other words, the call option on the coupon bond can be decomposed into a finite number of call options on zero-coupon bonds. In this case, if we assume the one-factor Gaussian model (6.22), the formula (6.37) applies. As a result, the price $\phi_f(t)$ of the call option on the coupon bond can be written as

$$\begin{aligned}
\phi_c(t) &= \sum_{s_i \geq s} C_i [P(t,s_i)N(d_1(t,s,s_i)) - K_i P(t,s)N(d_2(t,s,s_i))] \\
&+ F\,[P(t,T)N(d_1(t,s,s_j)) - K_j P(t,s)N(d_2(t,s,s_j))].
\end{aligned} \tag{6.48}$$

where $d_1(t,s,s_i)$ and $d_2(t,s,s_i)$ are given in (6.38) and (6.39), respectively, with $K = K_i$ and $T = s_i$.

The corresponding put option on the coupon bond can be obtained using (6.40):

$$\begin{aligned}
\phi_p(t) &= \sum_{s_i \geq s} C_i [-P(t,s_i)N(-d_1(t,s,s_i)) + K_i P(t,s)N(-d_2(t,s,s_i))] \\
&+ F\,[-P(t,T)N(-d_1(t,s,s_j)) + K_j P(t,s)N(-d_2(t,s,s_j))].
\end{aligned} \tag{6.49}$$

More applications of the forward risk adjusted measure will be discussed in Chapter 7.

6.5 BARRIER HITTING PROBABILITIES REVISITED

In this section, we consider some complex barrier hitting time distributions. We begin with the hitting time distribution of Brownian motion when the barrier is a non-horizontal straight line. This case is essentially the same as that in Example 4.2 and can be dealt with similarly. Our purpose in presenting the problem here is to illustrate the use of the Girsanov Theorem. We then use the Girsanov Theorem to derive the distribution of Brownian motion at a fixed time, given that its path has hit a non-horizontal barrier earlier. Last, we consider the case of two barriers, where we illustrate extensively how to use the Reflection Principle.

Single Non-Horizontal Barrier

Consider now Brownian motion with drift μ: $W_\mu(t) = \mu t + W(t)$, where $\{W(t)\}$ is standard Brownian motion under the probability measure P. Assume here without loss of generality that $\sigma = 1$. Let $x = a + kt$ be a straight line(horizontal or nonhorizontal). Define

$$T = \inf\{t; \ W_\mu(t) = a + kt\}. \tag{6.50}$$

Then T is the hitting time or the first passage time for the barrier $x = a + kt$. It is easy to see that

$$T = \inf\{t; \ W(t) - (k - \mu)t = a\}.$$

Let $\{\tilde{W}(t) = W(t) - (k - \mu)t\}$ so that $\{\tilde{W}(t)\}$ is Brownian motion with drift $k - \mu$. As shown in Section 4.3, the distribution of T is obtainable if $k - \mu = 0$. This suggests that if we are able to find a probability measure, say Q, such that $\{\tilde{W}(t)\}$ has zero drift under Q, we may use the result in Section 4.3 to obtain the distribution of T utilizing the relation between P and Q. In what follows, we will show that this approach is possible by applying the Girsanov Theorem.

Let

$$b = k - \mu \text{ and } Z_b(t) = e^{bW(t) - \frac{1}{2}b^2 t}. \tag{6.51}$$

Then by the Girsanov Theorem (Version A), $\{\tilde{W}(t) = W(t) - (k - \mu)t\}$ is standard Brownian motion under the probability measure Q generated by $Z_b(T)$ given in (6.51), and T is the first passage time of standard Brownian motion under Q. In this case the density $f_a(t)$ of T under Q is given in (4.10), i.e. $f_a(t) = \frac{|a|}{\sqrt{2\pi t^3}} e^{-\frac{a^2}{2t}}$. Let $f(t; a, k)$ be the density function of T under the original probability measure P. Since

$$Z_b^{-1}(t) = e^{-b\tilde{W}(t) - \frac{1}{2}b^2 t},$$

one has

$$
\begin{aligned}
f(t; a, k)dt &= E\{\mathbf{I}_{\{\mathcal{T} \in [t, t+dt)\}}\} \\
&= E_Q\{\mathbf{I}_{\{\mathcal{T} \in [t, t+dt)\}} Z_b^{-1}(t)\} = Z_b^{-1}(t) f_a(t) dt \\
&= e^{-ba - \frac{1}{2}b^2 t} \times \frac{|a|}{\sqrt{2\pi t^3}} e^{-\frac{a^2}{2t}} dt = \frac{|a|}{\sqrt{2\pi t^3}} e^{-\frac{(bt+a)^2}{2t}} dt.
\end{aligned}
$$

The second equality follows from the remark following the proof of the Girsanov Theorem, and the equality $Z_b^{-1}(t) = e^{-ba - \frac{1}{2}b^2 t}$ because $\mathcal{T} \in [t, t+dt)$ and $\tilde{W}(\mathcal{T}) = a$. Thus

$$f(t; a, k) = \frac{|a|}{\sqrt{2\pi t^3}} e^{-\frac{(bt+a)^2}{2t}}, \tag{6.52}$$

where b is given in (6.51).

To identify the distribution of $\mathcal{T}$, we compare it to the inverse Gaussian(IG) distribution given in Example 2.8. If a/b is negative, (6.52) is an inverse Gaussian density with $\alpha = -a/b$ and $\beta = 1/b^2$. If a/b is positive, (6.52) is not an inverse Gaussian density but can be expressed in terms of an inverse Gaussian density as follows:

$$f(t; a, k) = e^{-2ab} \left[\frac{|a|}{\sqrt{2\pi t^3}} e^{-\frac{(bt-a)^2}{2t}} \right]. \tag{6.53}$$

Noting that the second part is an inverse Gaussian density and the first part is a constant less than 1, $f(t; a, k)$ is a defective inverse Gaussian density.

Based on the above results, we may make the following conclusion. If $a/b < 0$, the distribution of the hitting time $\mathcal{T}$ defined in (6.50) for Brownian motion with drift μ is an inverse Gaussian distribution with $\alpha = -a/b$ and $\beta = 1/b^2$. Since the distribution is nondefective, $\mathcal{T}$ is a finite random variable. In this case, Brownian motion $\{\mu t + W(t)\}$ will eventually hit the barrier $x = a + kt$. If $a/b > 0$, the distribution is defective. The probability that the Brownian motion process hits the barrier is $\Pr\{\mathcal{T} < \infty\} = e^{-2ab}$. Further, given that the Brownian motion process hits the barrier, the hitting time distribution is inverse Gaussian with $\alpha = a/b$ and $\beta = 1/b^2$. $\qquad \square$

We next consider the distribution of the brownian motion $\{W_\mu(\mathcal{T})\}$ at time $\mathcal{T}$, given that it has hit the barrier $x = a + kt$ earlier. First let's assume $a > 0$. Let $g(x; a, k)$ be the corresponding density function. The derivation of this density is very similar to that of $f(x; a, k)$ given in (6.52). Continuing to use the notation for the non-horizontal barrier hitting time, we have

$$
\begin{aligned}
g(x; a, k)dx &= E\{\mathbf{I}_{\{W(T)\in[x,x+dx),W_\mu(t)=a+kt,\text{ for some }0\leq t\leq T\}}\} \\
&= E_Q\{\mathbf{I}_{\{\tilde{W}(T)\in[y,y+dy),\tilde{W}(t)=a,0\leq t\leq T\}}Z_b^{-1}(T)\} \\
&\quad (\text{where } y = x - (k - \mu)T = x - bT) \\
&= e^{-by-\frac{1}{2}b^2T}g_a(y)dy \\
&\quad (\text{since } Z_b^{-1}(T) = e^{-b\tilde{W}(T)-\frac{1}{2}b^2T} = e^{-by-\frac{1}{2}b^2T}) \\
&= e^{-bx+\frac{1}{2}b^2T}g_a(x - bT)dx,
\end{aligned}
$$

where $g_a(x)$ is given in (4.11) and is the density of standard Brownian motion at T, given that the Brownian motion process has hit the barrier earlier. We thus obtain

$$
g(x; a, k) = \begin{cases} \frac{1}{\sqrt{2\pi T}}e^{-\frac{(x-2a)^2}{2T}-2ab}, & x < a + bT \\ \frac{1}{\sqrt{2\pi T}}e^{-\frac{x^2}{2T}}, & x \geq a + bT. \end{cases} \tag{6.54}
$$

Similarly for $a < 0$, we have

$$
g(x; a, k) = \begin{cases} \frac{1}{\sqrt{2\pi T}}e^{-\frac{(x-2a)^2}{2T}-2ab}, & x > a + bT \\ \frac{1}{\sqrt{2\pi T}}e^{-\frac{x^2}{2T}}, & x \leq a + bT. \end{cases} \tag{6.55}
$$

The derivation of (6.55) is omitted. ☐

Double Barriers

We now consider the case of two horizontal barriers: one upper barrier and one lower barrier. We will derive the distribution of standard Brownian motion at a fixed time T given that it has hit at least one of the barriers earlier.

Let $\{W(t)\}$ be standard Brownian motion for $0 < t \leq T$, and let $x = a_u$ and $x = a_l$ be the upper and lower barriers, respectively, where $a_l < 0 < a_u$. Let $g(x; a_l, a_u)$ denote the density function of $W(T)$, given that $\{W(t)\}$ has hit at least one of the barriers earlier. In what follows, we derive an explicit expression for $g(x; a_l, a_u)$. Our main tool is the Reflection Principle introduced in Section 4.3.

We begin with two sequences of events defined as follows. For each positive integer n, let A_n be the event that there exist n times $0 < t_1 < t_2 < \cdots < t_n \leq T$ such that $W(t_{2k-1}) = a_u, W(t_{2k}) = a_l, k = 1, 2, \cdots$, and $W(T) \in [x, x + dx)$. In other words, A_n is the event that the Brownian motion process hits the barriers alternatively at least n times starting with the upper barrier, and takes a value in $[x, x+dx)$ at time T (this does not mean that the Brownian motion process hits the barriers exactly n times. Further it may hit the barriers between the times t_k). Similarly, we let B_n be the event that there exist n times $0 < t_1 < t_2 < \cdots < t_n \leq T$ such that $W(t_{2k-1}) = a_l, W(t_{2k}) = a_u, k = 1, 2, \cdots$, and $W(T) \in [x, x + dx)$. This is the event that the Brownian motion process hits the barriers alternatively at least n times starting with the lower barrier, and takes a value in $[x, x + dx)$ at time T. Observe that $A_{n-1} \cap B_{n-1} = A_n \cup B_n$. Since

$$P(A_{n-1} \cup B_{n-1}) = P(A_{n-1}) + P(B_{n-1}) - P(A_{n-1} \cap B_{n-1}),$$

we have

$$
\begin{aligned}
g(x; a_l, a_u)dx &= P(A_1 \cup B_1) = P(A_1) + \Pr(B_1) - P(A_1 \cap B_1) \\
&= P(A_1) + P(B_1) - P(A_2 \cup B_2) \\
&= P(A_1) + P(B_1) - P(A_2) - \Pr(B_2) + P(A_2 \cap B_2) \\
&\quad \cdots\cdots \\
&= \sum_{n=1}^{\infty} (-1)^{n-1}[P(A_n) + P(B_n)].
\end{aligned}
\tag{6.56}
$$

Thus, the problem reduces to the computation of the probabilities $P(A_n)$ and $P(B_n)$. It is easy to see that

$$P(A_1) = \frac{1}{\sqrt{2\pi T}} e^{-\frac{(x-2a_u)^2}{2T}} dx.$$

By applying the Reflection Principle twice, we have

$$P(A_2) = \frac{1}{\sqrt{2\pi T}} e^{-\frac{[x+2(a_u-a_l)]^2}{2T}} dx.$$

In general, we have

$$P(A_{2n-1}) = \frac{1}{\sqrt{2\pi T}} e^{-\frac{[x-2a_l-2n(a_u-a_l)]^2}{2T}} dx,$$

by applying the Reflection Principle $2n - 1$ times and

$$P(A_{2n}) = \frac{1}{\sqrt{2\pi T}} e^{-\frac{[x+2n(a_u-a_l)]^2}{2T}} dx,$$

by applying the Reflection Principle $2n$ times. Exchanging a_u and a_l in the above, we obtain

$$P(B_{2n-1}) = \frac{1}{\sqrt{2\pi T}} e^{-\frac{[x-2a_u-2n(a_u-a_l)]^2}{2T}} dx,$$

and

$$P(B_{2n}) = \frac{1}{\sqrt{2\pi T}} e^{-\frac{[x-2n(a_u-a_l)]^2}{2T}} dx.$$

Thus, from (6.56)

$$
\begin{aligned}
&g(x;\; a_l, a_u)dx \\
&= \sum_{n=1}^{\infty}(-1)^{n-1}[P(A_n) + P(B_n)] \\
&= \sum_{n=1}^{\infty}\{[P(A_{2n-1}) + P(B_{2n-1})] - [P(A_{2n}) + P(B_{2n})]\} \\
&= \sum_{n=1}^{\infty}\frac{1}{\sqrt{2\pi T}}\left[e^{-\frac{[x-2a_l-2n(a_u-a_l)]^2}{2T}} + e^{-\frac{[x-2a_u-2n(a_u-a_l)]^2}{2T}}\right]dx \\
&\quad- \sum_{n=1}^{\infty}\frac{1}{\sqrt{2\pi T}}\left[e^{-\frac{[x+2n(a_u-a_l)]^2}{2T}} + e^{-\frac{[x-2n(a_u-a_l)]^2}{2T}}\right]dx.
\end{aligned}
$$

Therefore,

$$
\begin{aligned}
g(x;\; a_l, a_u) &= \sum_{n=1}^{\infty}\frac{1}{\sqrt{2\pi T}}\left[e^{-\frac{[x-2a_l-2n(a_u-a_l)]^2}{2T}} + e^{-\frac{[x-2a_u-2n(a_u-a_l)]^2}{2T}}\right] \\
&\quad- \sum_{n=1}^{\infty}\frac{1}{\sqrt{2\pi T}}\left[e^{-\frac{[x+2n(a_u-a_l)]^2}{2T}} + e^{-\frac{[x-2n(a_u-a_l)]^2}{2T}}\right]. \quad (6.57)
\end{aligned}
$$

The distribution of $W(T)$, given that $\{W(t)\}$ has hit neither of the barriers earlier can now be derived easily. Let $h(x;\; a_l, a_u)$ be the corresponding density function. Then the relation between $h(x;\; a_l, a_u)$, and $g(x;\; a_l, a_u)$ is given by

$$h(x;\; a_l, a_u) = \frac{1}{\sqrt{2\pi T}} e^{-\frac{x^2}{2T}} - g(x;\; a_l, a_u),\; a_l < x < a_u.$$

Thus, $h(x;\; a_l, a_u)$ may be computed using Formula (6.57). If $x \leq a_l$ or $x \geq a_u$, one obviously has $h(x;\; a_l, a_u) = 0$.

Applying the Girsanov Theorem as in the single non-horizontal barrier case, we can calculate the density of Brownian motion with non-zero drift at time T with two parallel non-horizontal linear barriers. We leave the derivation to interested readers. □

6.6 TWO-DIMENSIONAL STOCHASTIC DIFFERENTIAL EQUATIONS

Two or higher-dimensional stochastic processes and stochastic differential equations are necessary when modelling more than one risky asset or modelling jointly risky assets and stochastic interest rates. In this section, we extend the results in Chapters 5 and 6 to two-dimensional stochastic differential equations. These results may further be extended to higher-dimensional stochastic differential equations There is no fundamental difference between the two-dimensional and higher-dimensional cases. Hence, we restrict ourselves to the two-dimensional case.

A pair of standard Brownian motion processes $\{(W_1(t), W_2(t))\}$ is said to be correlated two-dimensional standard Brownian motion if

1. increments $W_1(t) - W_1(s)$ and $W_2(t) - W_2(s), t > s$, are independent of $W_1(y)$ and $W_2(y)$ for any $0 \leq y \leq s$. In other words, the pair of processes as a vector has independent increments; and

2. the covariance

$$Cov(W_1(t), W_2(t)) = E\{W_1(t)W_2(t)\} = \rho t,$$

where $-1 \leq \rho \leq 1$.

Immediate implications are that the correlation coefficient of $W_1(t)$ and $W_2(t)$ is

$$Corr(W_1(t), W_2(t)) = \rho,$$

and for any $s \neq t$,

$$Cov(W_1(t), W_2(s)) = \rho \min(t, s).$$

If $\rho = 0$, the two components of two-dimensional standard Brownian motion are uncorrelated. In this case, $W_1(t)$ and $W_2(s)$ are independent for any t and s. We may express correlated two-dimensional Brownian motion in terms of uncorrelated two-dimensional standard Brownian motion. To see this, let $\{(W_1(t), W_2(t))\}$ be uncorrelated standard Brownian motion. Define

$$
\begin{aligned}
\tilde{W}_1(t) &= W_1(t), \\
\tilde{W}_2(t) &= \rho W_1(t) + \sqrt{1 - \rho^2} W_2(t).
\end{aligned}
\tag{6.58}
$$

$\{(\tilde{W}_1(t), \tilde{W}_2(t))\}$ is two-dimensional Brownian motion with correlation coefficient ρ. Two-dimensional Brownian motion with arbitrary drift and volatility can be introduced as a linear combination of two-dimensional standard Brownian motion processes similar to the one-dimensional case discussed in Section 4.2.

The martingale property for two-dimensional Brownian motion still holds. Similar to (4.27), $\{(W_1(t), W_2(t))\}$ is uncorrelated standard Brownian motion if and only if

$$Z_{\lambda_1, \lambda_2}(t) = e^{\lambda_1 W_1(t) + \lambda_2 W_2(t) - \frac{1}{2}[\lambda_1^2 + \lambda_2^2]t} \tag{6.59}$$

is a martingale (with respect to the natural information structure generated by $\{(W_1(t), W_2(t))\}$), for any pair of real values (λ_1, λ_2).

A pair of stochastic processes $\{(X_1(t), X_2(t))\}$ is a solution of a two-dimensional SDE

$$
\begin{aligned}
dX_1 &= \alpha_1(t, X_1, X_2)dt + \sigma_{11}(t, X_1, X_2))dW_1 + \sigma_{12}(t, X_1, X_2))dW_2, \\
dX_2 &= \alpha_2(t, X_1, X_2)dt + \sigma_{21}(t, X_1, X_2))dW_1 + \sigma_{22}(t, X_1, X_2))dW_2,
\end{aligned}
$$

(6.60)

where $\{(W_1(t), W_2(t))\}$ is uncorrelated standard Brownian motion, if

$$
\begin{aligned}
X_1(t) &= X_1(0) + \int_0^t \alpha_1(s, X_1(s), X_2(s))ds \\
&+ \sum_{j=1}^2 \int_0^t \sigma_{1j}(s, X_1(s), X_2(s))dW_j(s)
\end{aligned}
$$

and

$$
\begin{aligned}
X_2(t) &= X_2(0) + \int_0^t \alpha_2(s, X_1(s), X_2(s))ds \\
&+ \sum_{j=1}^2 \int_0^t \sigma_{2j}(s, X_1(s), X_2(s))dW_j(s).
\end{aligned}
$$

The conditions which guarantee the existence and uniqueness of a solution remain the same as that in the one-dimensional case, i.e. all $\alpha_i(t, x_1, x_2)$ and $\sigma_{ij}(t, x_1, x_2)$, $i = 1, 2$, $j = 1, 2$, satisfy the Lipschitz Condition and the Linear Growth Condition: there is a constant L such that

1. for any t, and any (x_1, x_2) and (y_1, y_2),

$$
|\alpha_i(t, x_1, x_2) - \alpha_i(t, y_1, y_2)| \leq L[|x_1 - y_1| + |x_2 - y_2|],
$$

   and

$$
|\sigma_{ij}(t, x_1, x_2) - \sigma_{ij}(t, y_1, y_2)| \leq L[|x_1 - y_1| + |x_2 - y_2|];
$$

2. for any t and x_1, x_2,

$$
|\alpha_i(t, x_1, x_2)| \leq L[1 + |x_1| + |x_2|],
$$

   and

$$
|\sigma_{ij}(t, x_1, x_2)| \leq L[1 + |x_1| + |x_2|].
$$

The following theorems are the analogs of Ito's Lemma, the exponential martingale and the Girsanov Theorem in the one-dimensional case.

Theorem 6.2 Ito's Lemma

Let $\{(X_1(t), X_2(t))\}$ be a solution to SDE (6.60) and $g(t, x_1, x_2)$ be a function which is continuously differentiable in t and continuously twice differentiable jointly with respect to x_1 and x_2. Then $g(t, X_1(t), X_2(t))$ is a solution of the following SDE

$$
\begin{aligned}
& dg(t, X_1, X_2) \\
= \;& \frac{\partial g(t, X_1, X_2)}{\partial t} dt + \frac{\partial g(t, X_1, X_2)}{\partial x_1} dX_1 + \frac{\partial g(t, X_1, X_2)}{\partial x_2} dX_2 \\
+ \;& \frac{1}{2} \left[\left(\sigma_{11}^2 + \sigma_{12}^2 \right) \frac{\partial^2 g(t, X_1, X_2)}{\partial x_1^2} + \left(\sigma_{21}^2 + \sigma_{22}^2 \right) \frac{\partial^2 g(t, X_1, X_2)}{\partial x_2^2} \right. \\
+ \;& \left. 2 \left(\sigma_{11}\sigma_{21} + \sigma_{12}\sigma_{22} \right) \frac{\partial^2 g(t, X_1, X_2)}{\partial x_1 \partial x_2} \right] dt.
\end{aligned}
$$

$$(6.61)$$

It can be seen that Theorem 5.2 is an immediate consequence of this theorem with $g(t, x_1, x_2) = x_1 x_2$.

Theorem 6.3 Exponential Martingale

Let $\{(X_1(t), X_2(t))\}$ be a solution to SDE (6.60). For any stochastic process $\{(b_1(t), b_2(t))\}$, if $b_i(t)\sigma_{ij}(t, X_1, X_2), i, j = 1, 2$, satisfies Novikov's Condition, then

$$
\begin{aligned}
Z_{b_1, b_2}(t) = \exp \Bigg\{ & \int_0^t [b_1\sigma_{11} dW_1(s) + b_1\sigma_{12} dW_2(s)] \\
+ & \int_0^t [b_2\sigma_{21} dW_1(s) + b_2\sigma_{22} dW_2(s)] \\
- & \frac{1}{2} \int_0^t (\sigma_{11}^2 + \sigma_{12}^2) b_1^2(s) ds \\
- & \int_0^t (\sigma_{11}\sigma_{21} + \sigma_{12}\sigma_{22}) b_1(s) b_2(s) ds - \frac{1}{2} \int_0^t (\sigma_{21}^2 + \sigma_{22}^2) b_2^2(s) ds \Bigg\}
\end{aligned}
$$

$$(6.62)$$

is a martingale.

Theorem 6.4 The Girsanov Theorem

Let $\{(X_1(t), X_2(t))\}$ be a solution to SDE (6.60). For given functions $\beta_1(t, x_1, x_2)$ and $\beta_2(t, x_1, x_2)$, suppose there are functions $\gamma_1(t, x_1, x_2)$ and $\gamma_2(t, x_1, x_2)$ for which

$$
\beta_1(t, x_1, x_2) = \alpha_1(t, x_1, x_2)
$$

$$\begin{aligned}
& + && \sigma_{11}(t, x_1, x_2)\gamma_1(t, x_1, x_2) + \sigma_{12}(t, x_1, x_2)\gamma_2(t, x_1, x_2) \\
\beta_2(t, x_1, x_2) & = && \alpha_2(t, x_1, x_2) \\
& + && \sigma_{21}(t, x_1, x_2)\gamma_1(t, x_1, x_2) + \sigma_{22}(t, x_1, x_2)\gamma_2(t, x_1, x_2),
\end{aligned}$$

$$(6.63)$$

and that $\gamma_1(t, X_1, X_2)$ and $\gamma_2(t, X_1, X_2)$ satisfy Novikov's Condition. Then, there exists an equivalent probability measure Q under which $\{(X_1(t), X_2(t))\}$ is a solution to the following SDE

$$\begin{aligned}
dX_1 &= \beta_1(t, X_1, X_2)dt + \sigma_{11}(t, X_1, X_2)d\tilde{W}_1 + \sigma_{12}(t, X_1, X_2)d\tilde{W}_2, \\
dX_2 &= \beta_2(t, X_1, X_2)dt + \sigma_{21}(t, X_1, X_2)d\tilde{W}_1 + \sigma_{22}(t, X_1, X_2)d\tilde{W}_2,
\end{aligned}$$

$$(6.64)$$

where $\{(\tilde{W}_1(t), \tilde{W}_2(t))\}$ is standard Brownian motion under Q.

CHAPTER 7

APPLICATIONS IN INSURANCE

In this chapter, we illustrate how stochastic calculus may be applied for the valuation of insurance and annuity products and and in particular of those whose benefits are linked to the performance of the bond and equity markets. In Section 7.1, we consider Deferred Variable Annuities (VAs) that are essentially selected mutual funds wrapped with guaranteed death and living benefits. Variable Annuities are sold by North American insurance companies as tax-deferred retirement saving products. The yearly sales of variable annuities has recently reached $126 billion and the total assets are more than $994 billion, according to the 2004 Annuity Fact Book published by the National Association for Variable Annuities (NAVA). A similar savings product called Equity-Indexed Annuity (EIA) is also considered in this section. EIA's are fixed annuities wrapped with guaranteed death and living benefits but they differ from Variable Annuities in terms of pricing method and asset management. Equity-Indexed Annuities were introduced in 1995 and have been gaining popularity in recent years. Sales were over $12 billion in 2003. For more details, see Lin and Tan (2003). Section 7.2 considers the valuation of guaranteed annuity options (GAOs). A guaranteed annuity option allows a pension policyholder to convert his or her account value into a life annuity at a fixed annuitization rate at retirement. Guaranteed annuity options have received considerable attentions from actuarial practitioners and researchers. See Boyle and Hardy (2003) and Ballotta and Haberman (2003). The last section considers the valuation of a Universal Life (UL) Insurance contract that allows its premiums to be invested in equity index funds. The model and methodology are due to Manistre (2001). A comprehensive discussion on embedded guarantees in insurance products can be found in Hardy (2003).

7.1   DEFERRED VARIABLE ANNUITIES AND EQUITY-INDEXED ANNUITIES

A Deferred Variable Annuity is a tax-deferred retirement saving plan that offers its holder to accumulate savings and defer taxes until the age of retirement. It has two phases: an accumulation phase and a payout phase. During the accumulation phase, premiums go into the holder's choice of selected stock and/or bond funds to accumulate assets. These funds are often called subaccounts in insurance terminology. The rate of return depends on the performance of the underlying subaccounts. The holder of a VA may convert the accumulated value at retirement into a lump-sum payment or into a fixed immediate life annuity. Variable Annuities are wrapped with a guaranteed minimum death benefit (GMDB). A GMDB provides a minimum level of protection against the financial loss from the investment of the subaccount at the time of death. The most basic guarantee is the so-called 'Return of Premium' guarantee under which the benefit level is the greater of (a) the subaccount value or (b) the total premiums paid in the past. Enhanced GMDBs are common but usually offered as a rider. Among them are the 'Annual Rachet' or annual 'Step Up' that guarantees the greater of (a), (b) mentioned earlier or (c) the account value on a prior contract anniversary date and the annual 'Roll Up' that guarantees a minimum rate of return for each year. A Variable Annuity often offers a guaranteed minimum living benefit (GMLB) as a rider. Depending on the delivery method of the benefit payouts, a GMLB can further be a guaranteed minimum accumulation benefit (GMAB) or a guaranteed minimum income benefit (GMIB)[1]. A basic GMAB guarantees a minimum surrender value at the end of the accumulation period and it is often a 'Return of Premium' or a 'Roll Up'. A basic GMIB gives the holder the right to annuitize at the end of the accumulation period. A minumum base is guaranteed for annuitization. To cover the costs associated with the management of the subaccount and the guarantees, fees (Fund Expenses and Mortality and Expenses Charges) proportional to the value of the subaccount are deducted from the subaccount.

[1] More recently, a new guarantee called the guaranteed minimum withdrawl benefit (GMWB) was introduced but we will not discuss it in this section.

Two cases will be discussed in this section. In the first case, the premiums of a Variable Annuity are invested in a bond fund. In the second case, the premiums are invested in an equity index fund.

Let δ be the proportional fee on a fund, and $F(t)$ and $F_\delta(t)$ be the values of the fund at time t without and with the fee. Without the loss of generality, assume $F(0) = 1$. Obviously, $F_\delta(t) = e^{-\delta t} F(t)$. Our objective is to determine the proportional fee δ.

We begin by considering a VA invested in a bond fund and with the Return of Premium guarantee. Assume again the one-factor Gaussian forward rate model (5.41) introduced in Section 5.4 for interest rates. That is, the forward rate $f(t, s)$ satisfies

$$df(t, s) = \alpha(t, s)dt + \sigma(t, s)\, dW(t),\ 0 \le t \le s, \tag{7.1}$$

where $W(t)$, $0 \le t \le s$, is standard Brownian motion, the volatility $\sigma(t, s)$ is deterministic, and

$$\alpha(t, s) = \sigma(t, s) \int_t^s \sigma(t, y)dy \tag{7.2}$$

under the risk-neutral measure Q.

Assume the value $F(t)$ of the bond fund satisfies

$$dF(t) = r(t)F(t)dt - \sigma_F(t)F(t)dW(t), \tag{7.3}$$

where $\sigma_F(t)$ is the fund volatility and is also determinstic. Let $D_\delta(s)$ be the payoff of the Return of Premium guarantee if it is paid at time s. Thus

$$D_\delta(s) = \max\{F_\delta(s),\ F_\delta(0)\}, \tag{7.4}$$

which may, in turn, be expressed as

$$D_\delta(s) = 1 + e^{-\delta s} \max\{F(s) - e^{\delta s},\ 0\}. \tag{7.5}$$

The payoff of this guarantee is the same of a call option on the fund and hence it may be valuated using the forward risk-adjusted measure in Section 6.4.

As shown in (6.32), the stochastic process $F(t)/P(t, s)$, $0 \le t \le s$, is a martingale under the forward risk adjusted measure Q_s, where $P(t, s)$ is the time-t price of the zero-coupon bond of par value 1 maturing at time s and the measure Q_s is defined in Section 6.4. Moreover, the process is lognormal with volatility

$$\sigma_F(t, s) = \int_t^s \sigma(t, y)dy - \sigma_F(t). \tag{7.6}$$

Similar to (6.37), the value of the payoff (7.4) at time 0 is given by

$$\phi_D(s) = E\left[e^{-\int_0^s r(u)du} D_\delta(s) \right]$$

$$
\begin{aligned}
&= P(0,s)\left\{1 + e^{-\delta s}E_s\{\max[F(s) - e^{\delta s},\, 0]\}\right\} \\
&= P(0,s)\left\{1 + e^{-\delta s}\left[\frac{1}{P(0,s)}N(d_1(s)) - e^{\delta s}N(d_2(s))\right]\right\} \\
&= P(0,s) + e^{-\delta s}N(d_1(s)) - P(0,s)N(d_2(s)) \\
&= e^{-\delta s}N(d_1(s)) + P(0,s)N(-d_2(s)),
\end{aligned}
\tag{7.7}
$$

where

$$
v(s)^2 = \int_0^s [\sigma_F(y,s)]^2 dy, \tag{7.8}
$$

$$
d_1(s) = \frac{-\delta s - \ln P(0,s) + \frac{1}{2}[v(s)]^2}{v(s)} \tag{7.9}
$$

and

$$
d_2(s) = d_1(s) - v(s). \tag{7.10}
$$

Several actuarial symbols are now provided in order to valuate the above and other benefit guarantees.

- (x): a life at age x;

- $T(x)$: the time of death or the future lifetime of (x);

- $_tq_x = P\{T(x) \le t\}$: the probability of death of (x) within t years after age x. $q_x = {_1q_x}$;

- $_tp_x = P\{T(x) > t\} = 1 - {_tq_x}$: the probability of survival of (x) for at least t years;

- $_tq_x^C$: the same as $_tq_x$ except that it is obtained from the Commissioners Extended Table (CET). In the CET, a positive loading is added to each q_x from the CSO table. These values are used to price death benefits;

- $_tp_x^C = 1 - {_tq_x^C}$;

- $_tq_x^A$: the same as $_tq_x$ except that it is obtained from an Annuity Table such as the GAM 80. In an Annuity Table, a percentage is deducted from each q_x in the CSO table. Survival benefits are often priced using an Annuity Table.

- $_tp_x^A = 1 - {_tq_x^A}$.

Suppose for simplicity that the benefit of the Return of Premium guarantee is payable at the end of the year of death. Let n be the year the variable annuity is converted into an immediate annuity. The value of the death benefit is

$$\sum_{k=0}^{n-1} \phi_D(k+1) \, _kp_x^C q_{x+k}^C$$

$$= \sum_{k=0}^{n-1} \left[e^{-\delta(k+1)} N(d_1(k+1)) + P(0, k+1)N(-d_2(k+1)) \right] \, _kp_x^C q_{x+k}^C,$$

which results in the following valuation formula

$$\sum_{k=0}^{n-1} \left[e^{-\delta(k+1)} N(d_1(k+1)) + P(0, k+1)N(-d_2(k+1)) \right] \, _kp_x^C q_{x+k}^C = 1.$$

$$(7.11)$$

The proportional fee δ can therefore be obtained by solving (7.11).

If, in addition, a Return of Premium accumulation benefit is offered, then the formula (7.11) may be modified to include it as follows:

$$\sum_{k=0}^{n-1} \left[e^{-\delta(k+1)} N(d_1(k+1)) + P(0, k+1)N(-d_2(k+1)) \right] \, _kp_x^C q_{x+k}^C$$

$$+ \left[e^{-\delta n} N(d_1(n)) + P(0, n)N(-d_2(n)) \right] \, _np_x^A = 1. \qquad (7.12)$$

Note that the survival probability from an Annuity Table is used for the accumulation benefit.

Suppose next that premiums of a Variable Annuity are invested in an equity index fund. In this case, we assume that the fund $F(t)$ is governed by

$$dF(t) = r(t)F(t)dt - \sigma_F(t)F(t)dW_F(t) \qquad (7.13)$$

under the risk-neutral probability measure Q, where $\sigma_F(t)$ is assumed to be deterministic, and $W_F(t)$ is a standard Brownian motion stochastically correlated with $W(t)$ in (7.1) via the correlation coefficient ρ. As shown in Section 6.6, we may write

$$W_F(t) = \rho W(t) + \sqrt{1 - \rho^2} W_f(t), \qquad (7.14)$$

where $W_f(t)$ is standard Brownian motion, independent of $W(t)$. The joint model (7.1) and (7.13) has been used for the valuation of Guaranteed Annuity Options by Boyle and Hardy (2003) and Ballotta and Haberman (2003). Their work will be discussed in the next section.

Again, we consider the Return of Premium guarantee (7.4). Under the same forward risk adjusted measure[2] Q_s,

$$dF(t) = [r(t) + \rho b(t, s)\sigma_F(t)] F(t)dt$$

---

[2]Since we are dealing with two-dimensinal Brownian motion, the two-dimensional Girsanov Theorem in Section 6.6 is used to derive the forward risk adjusted measure, even though the resulting probability measure remains the same.

$$- \rho \sigma_F(t) F(t) dW_s(t) - \sqrt{1 - \rho^2} \, \sigma_F(t) F(t) \, dW_f(t),$$

where $b(t, s)$ is given in (6.27). Apply the two-dimensional Ito's Lemma (6.61) to $F(t)$ and $P(t, s)$ with $g(F, P) = F/P$. Since

$$\frac{\partial g(F, P)}{\partial F} = \frac{1}{P}, \quad \frac{\partial g(F, P)}{\partial P} = -\frac{F}{P^2},$$

and

$$\frac{\partial^2 g(F, P)}{\partial^2 F} = 0, \quad \frac{\partial^2 g(F, P)}{\partial F \partial P} = -\frac{1}{P^2}, \quad \frac{\partial^2 g(F, P)}{\partial^2 P} = \frac{2}{P^3},$$

Ito's Lemma leads to

$$
\begin{aligned}
& d\frac{F(t)}{P(t, s)} \\
= \ & \frac{1}{P(t, s)} dF(t) - \frac{F(t)}{[P(t, s)]^2} dP(t, s) \\
+ \ & \frac{1}{2} \left\{ -2 \frac{1}{[P(t, s)]^2} [\rho \sigma_F(t) F(t)][b(t, s) P(t, s)] \right. \\
+ \ & \left. \frac{2}{[P(t, s)]^3} [b(t, s) P(t, s)]^2 \right\} dt \\
= \ & \frac{F(t)}{P(t, s)} \left\{ [r(t) + \rho b(t, s) \sigma_F(t)] dt \right. \\
+ \ & \left. \rho \sigma_F(t) dW_s(t) + \sqrt{1 - \rho^2} \, \sigma_F(t) dW_f(t) \right\} \\
- \ & \frac{F(t)}{P(t, s)} \left\{ [r(t) + [b(t, s)]^2 dt + b(t, s) dW_s(t) \right\} \\
+ \ & \frac{1}{2} \left\{ -2 \frac{1}{[P(t, s)]^2} [\rho \sigma_F(t) F(t)][b(t, s) P(t, s)] \right. \\
+ \ & \left. \frac{2}{[P(t, s)]^3} [b(t, s) P(t, s)]^2 \right\} dt \\
= \ & -\frac{F(t)}{P(t, s)} \left\{ [b(t, s) + \rho \sigma_F(t)] dW_s(t) + \sqrt{1 - \rho^2} \, \sigma_F(t) dW_f(t) \right\}.
\end{aligned}
$$
(7.15)

We again show that the process $F(t)/P(t, s)$, $0 \le t \le s$, is a martingale, and lognormally distributed for each fixed t. The log-variance of $F(t)/P(t, s)$ is given by

$$
\begin{aligned}
v^2(t, s) &= \int_0^t [b(y, s) + \rho \sigma_F(y)]^2 dy + \int_0^t [\sqrt{1 - \rho^2} \, \sigma_F(y)]^2 dy \\
&= \int_0^t [b^2(y, s) + 2\rho \sigma_F(y) b(y, s) + \sigma_F^2(y)] dy.
\end{aligned}
$$
(7.16)

Let

$$C(t, s) = E\left\{ e^{-\int_t^s r(u)du} \max[F(s) - e^{\delta s}, \, 0] \, \middle| \, \mathcal{F}_t \right\}.$$

The same argument as that for the derivation of (6.32) shows that the process $C(t, s)/P(t, s)$, $0 \le t \le s$, is a martingale. Hence, the value of the payoff (7.4) at time 0 can be expressed as

$$
\begin{aligned}
\phi_D(s) &= P(0, s)E_s\left[D_\delta(s)\right] \\
&= P(0, s)\left\{ 1 + e^{-\delta s} E_s\{\max[F(s) - e^{\delta s}, \, 0]\} \right\}. \quad (7.17)
\end{aligned}
$$

By applying the formula (5.63) with

$$\mu_X = 1/P(0, s) - \frac{1}{2}[v(s, s)]^2, \quad \sigma_X^2 = [v(s, s)]^2$$

to $E_s\{\max[F(s) - e^{\delta s}, \, 0]\}$, we obtain

$$
\begin{aligned}
\phi_D(s) &= P(0, s)\left\{ 1 + e^{-\delta s}\left[\frac{1}{P(0, s)}N(d_1(s)) - e^{\delta s}N(d_2(s)) \right] \right\} \\
&= e^{-\delta s}N(d_1(s)) + P(0, s)N(-d_2(s)), \quad (7.18)
\end{aligned}
$$

where

$$d_1(s) = \frac{-\delta s - \ln P(0, s) + \frac{1}{2}[v(s, s)]^2}{v(s, s)} \quad (7.19)$$

and

$$d_2(s) = d_1(s) - v(s, s). \quad (7.20)$$

The formulas (7.18), (7.19) and (7.20) are the same as (7.7), (7.9) and (7.10) except that $v(s)$ is replaced by $v(s, s)$. However, the difference is fundamental, as the former is completely driven by the random source $W(t)$ and the later is driven by the two random sources $W(t)$ and $W_f(t)$ as well as the correlation between them. With the closed form of the value of the guarantee payoff at each year derived, the rest of the valuation is exactly the same and the proportional fee can be computed via either (7.11) or (7.12), depending on whether a living benefit is included or not.

We now consider the valuation of an Equity-Indexed Annuity. Generally speaking, an Equity-Indexed Annuity or EIA is a hybrid of a traditional deferred fixed annuity and a Variable Annuity, but it is classified as a fixed annuity. The return of an EIA is linked to a pre-determined equity index such as S&P 500. It is also wrapped with a minimum return guarantee on a portion of the total premiums, which is required by nonforfeiture laws. An EIA has a fixed term, typically ranging from one to ten years. There are several ways to calculate the return of an EIA and they can loosely be described as Point-to-Point, Annual Reset, High-Water Mark, Annual Yield Spread, and Term

Yield Spread. Two important differences between an EIA and a VA are (i) that the premiums from an EIA go into the general account of its issuer and hence the actual return of the premium investment can be very different from the return of the equity index; and (ii) that the guarantee and the participation in the equity market are priced indirectly. An EIA is usually priced via the so-called participation rate that is roughly speaking a percentage of the index return.

The valuation of EIAs can be similar to the valuation of VAs. We only discuss the Point-to-Point design to illustrate. For other designs, their valuation and related practical issues, we refer readers to Lin and Tan (2003) and the references therein.

The Point-to-Point design is the simplest among all EIAs. Assume the initial premium of a Point-to-Point EIA to be one monetary unit. Under this design, the payoff of the EIA, payble at time s, can be expressed as

$$D_\alpha(s) = \max\{\min[1 + \alpha R(s),\, (1 + \zeta)^s],\, \beta(1 + \gamma)^s\}, \qquad (7.21)$$

where α is the participation rate yet to be determined, $R(s)$ is the total rate of return of an equity index between the period $[0, s]$, ζ is the maximum cap rate the EIA can earn, β is the premium base for calculating the minimum return guarantee and γ is the minimum interest rate. The above payoff structure is appealing to risk averse investors. While subject to the maximum cap rate that can be earned under this design, it allows investors to participate in any potential upside gain in the equity market. More importantly, in the event of an adverse market environment, the downside risk is constrained to the minimum guarantee floor component; i.e. $\beta(1 + \gamma)^s$. The presence of the cap rate is preferred by many insurers as it reduces the cost of such a design substantially.

Let $F(t)$, $F(0) = 1$, be the level of the equity index at time t and assume that the interest rates and the index follow the joint model (7.1) and (7.13). The total rate of return is $R(s) = F(s) - 1$. Thus the payoff $D_\alpha(s)$ given in (7.21) can be reformulated as the equivalent expression:

$$D_\alpha(s) = \beta(1+\gamma)^s + \alpha \max[F(s) - K_1,\, 0] - \alpha \max[F(s) - K_2,\, 0], \quad (7.22)$$

where

$$K_1 = \frac{1}{\alpha}[\beta(1 + \gamma)^s - 1 + \alpha]$$

and

$$K_2 = \frac{1}{\alpha}[(1 + \zeta)^s - 1 + \alpha].$$

The representation (7.22) is similar to (7.17) and hence allows for the derivation of the closed form for the value of the payoff $D_\alpha(s)$. Following the line of derivation for (7.18), we have

$$\phi_D(s) \;=\; \beta(1 + \gamma)^s P(0, s)$$

$$+ \quad \alpha P(0,s) \left[\frac{1}{P(0,s)} N(d_1(s)) - K_1 N(d_2(s)) \right]$$

$$- \quad \alpha P(0,s) \left[\frac{1}{P(0,s)} N(d_3(s)) - K_2 N(d_4(s)) \right]$$

$$= \quad \beta(1+\gamma)^s P(0,s) + \alpha N(d_1(s))$$

$$- \quad [\beta(1+\gamma)^s - 1 + \alpha] P(0,s) N(d_2(s))$$

$$- \quad \alpha N(d_3(s)) + [(1+\varsigma)^s - 1 + \alpha] P(0,s) N(d_4(s))$$

$$= \quad \alpha [N(d_1(s)) - N(d_3(s))]$$

$$+ \quad P(0,s) \{ \beta(1+\gamma)^s N(-d_2(s)) + (1-\alpha) N(d_2(s))$$

$$+ \quad [(1+\varsigma)^s - 1 + \alpha] N(d_4(s)) \}, \tag{7.23}$$

where

$$d_1(s) \quad = \quad \frac{\alpha - \ln[\beta(1+\gamma)^s - 1 + \alpha] - \ln P(0,s) + \frac{1}{2}[v(s,s)]^2}{v(s,s)},$$

$$d_2(s) \quad = \quad d_1(s) - v(s,s),$$

$$d_3(s) \quad = \quad \frac{\alpha - \ln[(1+\varsigma)^s - 1 + \alpha] - \ln P(0,s) + \frac{1}{2}[v(s,s)]^2}{v(s,s)},$$

$$d_4(s) \quad = \quad d_3(s) - v(s,s), \tag{7.24}$$

and $v(s,s)$ is given in (7.16). With the closed form (7.23), the participation rate α can be computed using the valuation formula

$$\sum_{k=0}^{n-1} \phi_D(k+1) \, _kp_x^C q_{x+k}^C + \phi_D(n) \, _np_x^A = 1, \tag{7.25}$$

where n is the term of the EIA.

The payoff representation (7.22) and the closed form (7.23) provide useful insights on hedging the EIA. The representation (7.22) suggests that one hedging approach is to construct a portfolio with the following three positions: (i) $\beta(1+\gamma)^s$ units of the zero-coupon bond maturing in year s, (ii) a long position of α units of call options with strike price K_1, and (iii) a short position of α units of call options with strike price K_2. A hedging strategy involving (ii) and (iii) is often referred to as a bull spread, a common practice in option trading (see Hull, 2000, pp. 187-189). The closed from suggests that an alternative hedging approach is to use the equity index and the zero-coupon bond directly with two positions: (i) $\alpha[N(d_1(s)) - N(d_3(s))]$ units of the index and (ii) $\beta(1+\gamma)^s N(-d_2(s)) + (1-\alpha) N(d_2(s)) + [(1+\varsigma)^s - 1 + \alpha] N(d_4(s))$ units of the zero-coupon bond.

Other EIA designs often have a path-dependent payoff structure and must be valuated numerically. For instance, an Annual Reset design has the payoff function:

$$D_\alpha(s) = \max\left\{ \prod_{i=1}^{s} \max\left[\min[1 + \alpha R(i), 1 + \varsigma], 1 \right], \beta(1 + \gamma)^s \right\},$$

where $R(i)$ and other parameters are defined similarly to those in the Point-to-Point design (7.21). Due to the path-dependent nature of the payoff function, no closed form for its value is available.

7.2 GUARANTEED ANNUITY OPTIONS

Many pension policies are wrapped with a guarantee under which the policyholders have a right to convert the account value of the policies into a life annuity at a fixed annuitization rate at retirement. As described in Boyle and Hardy (2003), these guarantees are often long term guarantees and can be very valuable under certain circumstances. As a result, the liabilites arising from these guarantees might have a serious financial impact on an insurance company with a significant exposure to these guarantees. A good example is the Equitable Life (UK) that offered GAOs in 1970's and 1980's at no cost and the accumulated liabilities from these guarantees partly led to the closure of the company to new business a few years ago. In this section, we use the model (7.1) and (7.13) to derive the closed form expression for the value of GAOs. The approaches are adopted from Boyle and Hardy (2003) and Pelsser (2003).

Let $F(t)$ be the account value, at time t, of a policy that offers a guaranteed annuity option with annuitization rate of g. That is, if an annuitant pays g dollars, he or she will receive one dollar at the beginning of each year, contingent on survival. Let also n be the time to retirement or the time to maturity and $\ddot{a}_x(n)$ be the fair value of a life annuity-due of one dollar issued at time n to a life who is at age x currently ($t = 0$). Thus $\ddot{a}_x(n)$ can be expressed as

$$\ddot{a}_x(n) = \sum_{k=n}^{\psi-x-1} P(n,k) \, _kp_x^A, \qquad (7.26)$$

where ψ is the maximum age of a life and $_kp_x^A$ is defined in Section 7.1. The payoff of the GAO at maturity time n, if it is exercised, is

$$\frac{F(n)}{g}\ddot{a}_x(n) - F(n) = F(n)\left(\frac{\ddot{a}_x(n)}{g} - 1 \right).$$

Hence, the payoff of the guarantee at maturity time n is

$$F(n) \max\left[\left(\frac{\ddot{a}_x(n)}{g} - 1 \right), 0 \right]. \qquad (7.27)$$

The payoff (7.27) resembles the payoff of the call option on a coupon bond (6.46) except that 'the number of units' of the bond is a random variable. The value of the guarantee at time t is

$$\phi_g = E\left\{e^{-\int_0^n r(u)du}F(n)\max\left[\left(\frac{\ddot{a}_x(n)}{g}-1\right),\,0\right]\right\}. \qquad (7.28)$$

To derive the closed form for the value of the guarantee at time 0, we further assume the volatility structure (6.41) of interest rates in Section 6.4. This assumption implies that the bond prices $P(n,k)$, $k = n, n+1, \cdots, \psi - x - 1$ have the affine form (6.43). Similar to the valuation of call options on coupon bonds, let r^* be the solution to the equation

$$\sum_{k=n}^{\psi-x-1} h(r,n,k)\ _k p_x^A = g,$$

where $h(r,n,k)$ is given in (6.45). Let $L_k = h(r^*,n,k), k = n, n + 1, \cdots, \psi - x - 1$. Then,

$$\max\left[\left(\frac{\ddot{a}_x(n)}{g}-1\right),\,0\right] = \sum_{k=n}^{\psi-x-1}\max[P(n,k)-L_k,\,0]\left[_k p_x^A/g\right].$$

Thus the expression (7.28) is rewritten as

$$\phi_g = \sum_{k=n}^{\psi-x-1} E\left\{e^{-\int_0^n r(u)du}F(n)\max[P(n,k)-L_k,\,0]\right\}\left[_k p_x^A/g\right].$$

$$(7.29)$$

We first focus on solving the expectation

$$E\left\{e^{-\int_0^n r(u)du}F(n)\max[P(n,k)-L_k,\,0]\right\}.$$

Under the forward risk adjusted measure Q_n, we have

$$E\left\{e^{-\int_0^n r(u)du}F(n)\max[P(n,k)-L_k,\,0]\right\}$$
$$= P(0,n)E_n\left\{F(n)\max[P(n,k)-L_k,\,0]\right\}. \qquad (7.30)$$

As shown in (6.35) and (7.15), the processes $P(t,k)/P(t,n)$, $0 \le t \le n$, and $F(t)/P(t,n)$, $0 \le t \le n$, satisfy the SDEs

$$d\frac{P(t,k)}{P(t,n)} = -\frac{P(t,k)}{P(t,n)}\left[\int_n^k \sigma(t,y)dy\right]dW_s(t)$$

and

$$
\begin{aligned}
d\frac{F(t)}{P(t,n)} &= -\frac{F(t)}{P(t,n)} \left\{ \left[\rho\sigma_F(t) - \int_t^n \sigma(t,y)dy \right] dW_s(t) \right. \\
&\quad + \left. \sqrt{1-\rho^2}\, \sigma_F(t)dW_f(t) \right\},
\end{aligned}
$$

respectively. For notational similicity, denote

$$
\begin{aligned}
\sigma_1(t) &= \int_n^k \sigma(t,y)dy, \\
\sigma_2(t) &= \rho\sigma_F(t) - \int_t^n \sigma(t,y)dy \\
&\text{and} \\
\sigma_3(t) &= \sqrt{1-\rho^2}\, \sigma_F(t).
\end{aligned} \tag{7.31}
$$

It follows from Ito's Lemma that

$$
\frac{P(t,k)}{P(t,n)} = \frac{P(0,k)}{P(0,n)} e^{-\frac{1}{2}\int_0^t [\sigma_1(u)]^2 du - \int_0^t \sigma_1(u)dW_s(u)}
$$

and

$$
\frac{F(t)}{P(t,n)} = \frac{1}{P(0,n)} e^{-\frac{1}{2}\int_0^t \left\{ [\sigma_2(u)]^2 + [\sigma_3(u)]^2 \right\} du - \int_0^t \sigma_2(u)dW_s(u) - \int_0^t \sigma_3(u)dW_f(u)}.
$$

Let $t = n$ and we have

$$
P(n,k) = \frac{P(0,k)}{P(0,n)} e^{-\frac{1}{2}\int_0^n [\sigma_1(u)]^2 du - \int_0^n \sigma_1(u)dW_s(u)}
$$

and

$$
F(n) = \frac{1}{P(0,n)} e^{-\frac{1}{2}\int_0^n \left\{ [\sigma_2(u)]^2 + [\sigma_3(u)]^2 \right\} du - \int_0^n \sigma_2(u)dW_s(u) - \int_0^t \sigma_3(u)dW_f(u)}.
$$

Introduce the normal random variables:

$$
X = \ln\frac{P(0,k)}{P(0,n)} - \frac{1}{2}\int_0^n [\sigma_1(u)]^2 du - \int_0^n \sigma_1(u)dW_s(u)
$$

and

$$
\begin{aligned}
Y &= -\frac{1}{2}\int_0^n \left\{ [\sigma_2(u)]^2 du + [\sigma_3(u)]^2 \right\} du \\
&\quad - \int_0^n \sigma_2(u)dW_s(u) - \int_0^t \sigma_3(u)dW_f(u).
\end{aligned}
$$

Then,

$$P(n, k) = e^X, \quad F(n) = \frac{1}{P(0, n)} e^Y.$$

The martingale property of Ito integrals gives

$$\mu_X = \ln \frac{P(0, k)}{P(0, n)} - \frac{1}{2} \int_0^n [\sigma_1(u)]^2 du$$

and

$$\mu_Y = -\frac{1}{2} \int_0^n \left\{ [\sigma_2(u)]^2 du + [\sigma_3(u)]^2 \right\} du.$$

Moreover, their variances are

$$\sigma_X^2 = \int_0^n [\sigma_1(u)]^2 du$$

and

$$\sigma_Y^2 = \int_0^n \left\{ [\sigma_2(u)]^2 du + [\sigma_3(u)]^2 \right\} du.$$

Since $W_s(u)$, $0 \le u \le n$, and $W_f(u)$, $0 \le u \le n$, are stochastically independent, the covariance is given by

$$\sigma_{XY} = \int_0^n [\sigma_1(u)\sigma_2(u)] du.$$

The expectation (7.30) can thus be computed using (5.64):

$$E \left\{ e^{-\int_0^n r(u)du} F(n) \max[P(n, k) - L_k, 0] \right\}$$
$$= P(0, n) E_n \left\{ F(n) \max[P(n, k) - L_k, 0] \right\}$$
$$= E_n \left\{ e^Y \max[e^X - L_k, 0] \right\}$$
$$= e^{\mu_X + \mu_Y + \frac{1}{2}(\sigma_X^2 + 2\sigma_{XY} + \sigma_Y^2)} N \left(\frac{\mu_X + \sigma_{XY} + \sigma_X^2 - \ln L_k}{\sigma_X} \right)$$
$$- L_k e^{\mu_Y + \frac{1}{2}\sigma_Y^2} N \left(\frac{\mu_X + \sigma_{XY} - \ln L_k}{\sigma_X} \right).$$

Some simple algebra leads to

$$e^{\mu_Y + \frac{1}{2}\sigma_Y^2} = 1,$$

$$e^{\mu_X + \mu_Y + \frac{1}{2}(\sigma_X^2 + 2\sigma_{XY} + \sigma_Y^2)} = \frac{P(0, k)}{P(0, n)} e^{\int_0^n [\sigma_1(u)\sigma_2(u)] du},$$

$$\frac{\mu_X + \sigma_{XY} + \sigma_X^2 - \ln L_k}{\sigma_X}$$
$$= \frac{\ln \frac{P(0,k)}{P(0,n)} + \int_0^n [\sigma_1(u)\sigma_2(u)] du + \frac{1}{2} \int_0^n [\sigma_1(u)]^2 du - \ln L_k}{\int_0^n [\sigma_1(u)]^2 du}$$

and

$$\frac{\mu_X + \sigma_{XY} - \ln L_k}{\sigma_X}$$

$$= \frac{\ln \frac{P(0,k)}{P(0,n)} + \int_0^n [\sigma_1(u)\sigma_2(u)]du - \frac{1}{2}\int_0^n [\sigma_1(u)]^2 du - \ln L_k}{\int_0^n [\sigma_1(u)]^2 du}.$$

Therefore,

$$E\left\{ e^{-\int_0^n r(u)du} F(n) \max[P(n,k) - L_k, \, 0] \right\}$$

$$= \frac{P(0,k)}{P(0,n)} e^{\int_0^n [\sigma_1(u)\sigma_2(u)]du} N(d_{1k}) - L_k N(d_{2k}), \qquad (7.32)$$

where

$$d_{1k} = \frac{\ln \frac{P(0,k)}{P(0,n)} + \int_0^n [\sigma_1(u)\sigma_2(u)]du + \frac{1}{2}\int_0^n [\sigma_1(u)]^2 du - \ln L_k}{\int_0^n [\sigma_1(u)]^2 du} \qquad (7.33)$$

and

$$d_{2k} = \frac{\ln \frac{P(0,k)}{P(0,n)} + \int_0^n [\sigma_1(u)\sigma_2(u)]du - \frac{1}{2}\int_0^n [\sigma_1(u)]^2 du - \ln L_k}{\int_0^n [\sigma_1(u)]^2 du}. \qquad (7.34)$$

Here, $\sigma_1(u)$ and $\sigma_2(u)$ are given in (7.31). Finally, the value of the guaranteed annuity option can be obtained as

$$\phi_g = \sum_{k=n}^{\psi-x-1} \left\{ \frac{P(0,k)}{P(0,n)} e^{\int_0^n [\sigma_1(u)\sigma_2(u)]du} N(d_{1k}) - L_k N(d_{2k}) \right\} \left[{}_k p_x^A / g \right],$$

$$(7.35)$$

where d_{1k} and d_{2k} are given in (7.33) and (7.34), respectively.

7.3 UNIVERSAL LIFE

Universal Life insurance may be viewed as a variation of Whole Life insurance or a term insurance with a side fund. A particularly attractive feature of Universal Life is that it allows a large portion of premiums to be invested in equity index funds. As a result, the cash value of Universal Life has a potential to grow faster than that of Whole Life, which may in turn increase death benefits over time under normal circumstances. Other attractive features of Universal Life include flexible premium payments, allowing early surrenders and the taking of policy loans. Because of the investment link to the financial markets and those flexibilities, Universal Life could be more risky than Whole Life in the sense that in an unfavorable economic environment there may not be sufficient funds in its account to cover the projected cost of the death benefits and its holders may be forced to increase premium payments.

In the following, a stochastic model for a Universal Life is presented. The model is taken from Manistre (2001) and is simplified for illustration purposes. Assume that a Universal Life is issued to (x) with maturity of n ($n = 100 - x$ for example). In order to precisely characterize the Universal Life under consideration, the following notation is introduced.

$S(t)$: the value of an equity index fund at time t;

$F(t)$: the account value of the Universal Life at time t to which the insurer credits premiums and interest;

$B(t)$: the accrued bonus at time t. We assume that bonuses are payable on anniversary dates;

$V(t)$: the reserve at time t, assumed to be a function of the values of R and B, i.e., $V(t) = V(t, R, B)$;

$\mu^d(t)$: the force of mortality at time t for (x);

$\mu^w(t)$: the force of surrender (or complete withdrawal) at time t for (x). At the time of surrender, a surrender charge $D(t)$ is imposed which we assume to be proportional to the account value, i.e., $D(t) = hF(t)$ with $h < 1$. We further assume for simplicity that there is no partial withdrawal.

We first model the index fund, the account value and the accrued bonus. Similar to the Black-Scholes framework (5.48), we assume that the index fund follows a geometric Brownian motion:

$$dS(t) = \alpha_S S(t)dt + \sigma_S S(t)dW(t), \qquad (7.36)$$

where α_S and σ_S are the drift and volatility of the index fund and $W(t)$ a standard Brownian motion. The stochastic differential equation for $F(t)$ is given by

$$dF(t) = F(t)\left[\frac{dS(t)}{S(t)} - \gamma dt\right] - c[F_0 + \epsilon F(t) - F(t)]dt + pdt, \quad (7.37)$$

where γ is the spread between the index fund return and the credited return to the insurance account, $F_0 + \epsilon F(t)$ is the death benefit where F_0 is the fixed portion and $\epsilon F(t)$ the variable portion proportional to the account value with $0 < \epsilon \leq 1$, c is the rate of Cost of Insurance (COI), and p is the annual premium payable continuously. Thus the contributions to the increment in account value are the credited returns and the premium income minus the cost of insurance. With (7.36), the equation (7.37) may be simplfied as

$$dF(t) = [\alpha_F F(t) + \beta_F] dt + \sigma_S F(t) dW(t), \qquad (7.38)$$

where $\alpha_F = \alpha_S - \gamma + c(1 - \epsilon)$ and $\beta_F = p - cF_0$. This is a linear stochastic differential equation as discussed in (5.77) of Chapter 5 and can be solved. The accrued bonus is determined based on the account value, the credited return and the cost of insurance and is expressed as

$$dB(t) = \phi_1 F(t)dt + \phi_2 F(t) \left[\frac{dS(t)}{S(t)} - \gamma dt \right] + \phi_3 c[F_0 + \epsilon F(t) - F(t)]dt,$$
$$(7.39)$$

with $B(0) = 0$, where the ϕ's are the weights. Rewriting (7.39) yields

$$dB(t) = [\alpha_B F(t) + \beta_B] dt + \phi_2 \sigma_S F(t) dW(t), \qquad (7.40)$$

where $\alpha_B = \phi_1 + \phi_2(\alpha_S - \gamma) - \phi_3 c(1 - \epsilon)$ and $\beta_B = \phi_3 cF$. As the bonus is credited and reset to zero on anniversary dates, we have that for $j = 1, 2, \cdots, n - 1$, $B(j) = 0$, $F(j) = F(j-) + B(j-)$, where $f(j-) = \lim_{t \uparrow j} f(t)$. Thus $F(t)$ and $B(t)$ are piece-wise stochastic processes governed by (7.38) and (7.40) with initial values being reset at $j = 1, 2, \cdots, n - 1$.

We next calculate the increment of the expected liability, $dL(t)$, given that the account value is $F(t)$ and (x) is alive at time t. The increment has three components: the death benefit in the case of death, the cost due to a surrender and the increment in reserve. Since the value of the death benefit is $F_0 + \epsilon F(t)$, the net cost for the death is $F_0 + \epsilon F(t) - V(t)$. Thus the expected cost is

$$[F_0 + \epsilon F(t) - V(t)]\mu^d(t)dt. \qquad (7.41)$$

Similarly, the expected cost for a surrender is

$$[F(t) - D(t) - V(t)]\mu^w(t)dt = [(1 - h)F(t) - V(t)]\mu^w(t)dt. \qquad (7.42)$$

Furthermore, the expected increase in liability excluding death and surrender is

$$[1 - \mu^d(t)dt - \mu^w(t)dt]dV(t). \qquad (7.43)$$

Thus we have

$$dL(t) = [F_0 + \epsilon F(t) - V(t)]\mu^d(t)dt + [(1 - h)F(t) - V(t)]\mu^w(t)dt + dV(t),$$
$$(7.44)$$

since $dtdV = 0$ as in Table 5.1.

To hedge against the liability, the reserve $V(t)$ is invested in the index fund $S(t)$ and a risk free asset that earns interest at rate r. Let $\theta(t)$ be the amount invested in the index fund. Then, the increment in the asset is

$$\begin{aligned} dA(t) &= [V(t) - \theta(t)]rdt + \theta(t)\frac{dS(t)}{S(t)} \\ &= [V(t) - \theta(t)]rdt + \theta(t)[\alpha_S dt + \sigma_S dW]. \end{aligned} \qquad (7.45)$$

as in (5.49). Matching the increments in the liability and the asset yields

$$[F_0 \ +\epsilon F(t) - V(t)]\mu^d(t)dt + [(1-h)F(t) - V(t)]\mu^w(t)dt + dV(t)$$
$$= [V(t) - \theta(t)]rdt + \theta(t)[\alpha_S dt + \sigma_S dW]. \tag{7.46}$$

We now obtain an equation similar to (5.50) that leads to the Black-Scholes Option Pricing Formula. Hence the methodology used in the derivation of the Black-Scholes Formula in Chapter 5 is applicable. As in Subsection 5.5, we apply the two-dimensional Ito's Lemma (Chapter 6, Theorem 6.2) to dV and obtain

$$
\begin{aligned}
dV &= \left[\frac{\partial V}{\partial t} + \frac{1}{2}\sigma_S^2 F^2 \left(\frac{\partial^2 V}{\partial F^2} + 2\phi_2 \frac{\partial^2 V}{\partial F \partial B} + \phi_2^2 \frac{\partial^2 V}{\partial B^2}\right)\right] dt \\
&+ \frac{\partial V}{\partial F}dF + \frac{\partial V}{\partial B}dB \\
&= \left[\frac{\partial V}{\partial t} + \frac{1}{2}\sigma_S^2 F^2 \left(\frac{\partial^2 V}{\partial F^2} + 2\phi_2 \frac{\partial^2 V}{\partial F \partial B} + \phi_2^2 \frac{\partial^2 V}{\partial B^2}\right)\right] dt \\
&+ \frac{\partial V}{\partial F}[(\alpha_F F + \beta_F)dt + \sigma_S F dW] \\
&+ \frac{\partial V}{\partial B}[(\alpha_B F + \beta_B)dt + \phi_2 \sigma_S F dW] \\
&= \left[\frac{\partial V}{\partial t} + (\alpha_F F + \beta_F)\frac{\partial V}{\partial F} + (\alpha_B F + \beta_B)\frac{\partial V}{\partial B}\right. \\
&+ \left.\frac{1}{2}\sigma_S^2 F^2 \left(\frac{\partial^2 V}{\partial F^2} + 2\phi_2 \frac{\partial^2 V}{\partial F \partial B} + \phi_2^2 \frac{\partial^2 V}{\partial B^2}\right)\right] dt \\
&+ \sigma_S F \left(\frac{\partial V}{\partial F} + \phi_2 \frac{\partial V}{\partial B}\right) dW. \tag{7.47}
\end{aligned}
$$

It follows from (7.46) that[3]

$$\theta = F\left(\frac{\partial V}{\partial F} + \phi_2 \frac{\partial V}{\partial B}\right) \tag{7.48}$$

and

$$
\begin{aligned}
&(V - \theta)r + \theta\alpha_S) \\
&= [F_0 + \epsilon F(t) - V(t)]\mu^d(t) + [(1-h)F(t) - V(t)]\mu^w(t) \\
&+ \frac{\partial V}{\partial t} + (\alpha_F F + \beta_F)\frac{\partial V}{\partial F} + (\alpha_B F + \beta_B)\frac{\partial V}{\partial B} \\
&+ \frac{1}{2}\sigma_S^2 F^2 \left(\frac{\partial^2 V}{\partial F^2} + 2\phi_2 \frac{\partial^2 V}{\partial F \partial B} + \phi_2^2 \frac{\partial^2 V}{\partial B^2}\right). \tag{7.49}
\end{aligned}
$$

---

[3]It should be understood that functions and processes such as θ, V, F, etc. are functions of t, but for expressional simplicity t is suppressed.

Substituting (7.48) into (7.49) yields

$$\frac{\partial V}{\partial t} + (\tilde{\alpha}_F F + \beta_F)\frac{\partial V}{\partial F} + (\tilde{\alpha}_B F + \beta_B)\frac{\partial V}{\partial B}$$
$$- (r + \mu^d + \mu^w)V + (\epsilon\mu^d + (1 - h)\mu^w)F + \mu^d F_0 = 0,) \tag{7.50}$$

where $\tilde{\alpha}_F = r - \gamma + c(1 - \epsilon)$ and $\tilde{\alpha}_B = \phi_1 + \phi_2(r - \gamma) - \phi_3 c(1 - \epsilon)$. We note that the PDE (7.50) does not contain the drift α_S.

Equation (7.50) suggests that the reserve $V(t)$ is of the form

$$V(t) = a(t)F(t) + b(t)B(t) + k(t), \tag{7.51}$$

where $a(t), b(t)$ and $k(t)$ are piece-wise deterministic functions and independent of R and B. In the following, we identify the functions $a(t), b(t)$ and $k(t)$. Substituting (7.51) into (7.50), we have

$$a'(t)F + b'(t)B + k'(t) + (\tilde{\alpha}_F F + \beta_F)a(t) + (\tilde{\alpha}_B F + \beta_B)b(t)$$
$$- (r + \mu^d + \mu^w)[a(t)F + b(t)B + k(t)]$$
$$+ (\epsilon\mu^d + (1 - h)\mu^w)F + \mu^d F_0 = 0.$$

Since the functions $a(t), b(t)$ and $k(t)$ are independent of F and B, the coefficients of F and B must be zero. We therefore obtain a system of three linear ordinary differential equations:

$$a'(t) = [r + \mu^d + \mu^w - \tilde{\alpha}_F]a(t) - \tilde{\alpha}_B b(t) - [\epsilon\mu^d + (1 - h)\mu^w], \tag{7.52}$$

$$b'(t) = [r + \mu^d + \mu^w]b(t), \tag{7.53}$$

$$k'(t) = -\beta_F a(t) - \beta_B b(t) + [r + \mu^d + \mu^w]k(t) - \mu^d F_0. \tag{7.54}$$

To solve (7.52) to (7.54) for each interval $[j - 1, j)$, $j = 1, 2, \cdots, n$, we need to identify a boundary condition of these equations. We begin with the last interval $[n - 1, n)$. Since at maturity the policy pays the face amount F_0 and the account value F (i.e., ϵ is set to be one), plus any accrued bonus, we have $V(n-) = F_0 + F + B$. Thus, $a(n-) = 1, b(n-) = 1$ and $k(n-) = F_0$. That is, for $n - 1 \le t < n$ the equations (7.52)-(7.54) have the terminal condition

$$a(n-) = 1, \quad b(n-) = 1 \text{ and } k(n-) = F_0. \tag{7.55}$$

For the interval $j - 1 \le t < j$, $j < n$, the continuity of the reserve process $V(t)$ implies that

$$a(j-)F(j-) + b(j-)B(j-) + k(j-) = a(j)F(j) + k(j)$$
$$= a(j)[F(j-) + B(j-)] + k(j).$$

Thus, the terminal conditions for $j - 1 \leq t < j$, $j = 1, 2, \cdots, n - 1$, are

$$a(j-) = b(j-) = a(j) \text{ and } k(j-) = k(j). \tag{7.56}$$

We can now solve (7.53)-(7.54) recursively in the order of $j = T, T-1, \cdots, 1$. It is obvious from the structure of these equations that we first solve for $b(t)$, and then for $a(t)$ and $k(t)$. With (7.56), it is easy to see

$$b(t) = a(j)e^{-\int_t^j [r + \mu^d(s) + \mu^w(s)]ds}, \quad j - 1 \leq t < j. \tag{7.57}$$

Write

$$l_a(t) = r + \mu^d(t) + \mu^w(t) - \tilde{\alpha}_F$$

and

$$m_a(t) = -\tilde{\alpha}_B b(t) - \epsilon \mu^d(t) - (1 - h)\mu^w(t).$$

Then (7.53) may be reexpressed as $a'(t) = l_a(t)a(t) + m_a(t)$, a first-order linear ODE. Thus, its solution with the terminal condition (7.56) is given by

$$a(t) = a(j)e^{-\int_t^j l_a(u)du} - \int_t^j m_a(s)e^{-\int_t^s l_a(u)du}ds, \quad j - 1 \leq t < j. \tag{7.58}$$

Similarly, write

$$l_k(t) = r + \mu^d(t) + \mu^w(t)$$

and

$$m_k(t) = -\beta_F a(t) - \beta_B b(t) - \mu^d(t)F_0.$$

We have

$$k(t) = k(j)e^{-\int_t^j l_k(u)du} - \int_t^j m_k(s)e^{-\int_t^s l_k(u)du}ds, \quad j - 1 \leq t < j. \tag{7.59}$$

Thus, we obtain a closed form expression for $a(t), b(t)$ and $k(t)$. The reserve process $V(t)$ then can be derived easily using (7.51) and the closed form solution for $F(t)$ and $B(t)$. The functions $a(t), b(t)$ and $k(t)$ are referred to in Manistre (2001) as the hedge ratio, bonus ratio and the insurance reserve. Finally, from $\theta = F\left(\frac{\partial V}{\partial F} + \phi_2 \frac{\partial V}{\partial B}\right)$, we have that at time t, the amount invested in the index fund is

$$\theta(t) = [a(t) + \phi_2 b(t)] F(t), \tag{7.60}$$

and the amount in the risk-free asset is

$$V(t) - \theta(t) = b(t)[B(t) - \phi_2 F(t)] + k(t). \tag{7.61}$$

Detailed discussions on insurance implications may be found in Manistre (2001) and are omitted.

REFERENCES

1. Abramowitz, M. (2002). *Handbook of Mathematical Functions with Formulas, Graphs, and Mathematical Tables.* Dover, New York.

2. Ballotta, L. and S. Haberman (2003). Valuation of guaranteed annuity conversion options, *Insurance, Mathematics and Economics,* 33, 87-108.

3. Benninga, S. (1997). *Financial Modeling,* MIT Press, Cambridge, MA.

4. Black F. and M.J. Scholes (1973). The pricing of options and corporate liabilities. *Journal of Political Economy,* 81, 637-654.

5. Black F., E. Derman and W. Toy (1990). A one factor model of interest rates and its application to treasury bond options, *Financial Analysts Journal,* 46, 33-39.

6. Björk, T. (1998). *Arbitrage Theory in Continuous Time,* Oxford University Press, New York.

7. Bolton, M.J. (1997). Reserving for annuity guarantees, *Report of the Annuity Guarantees Working Party,* Institute of Actuaries, London.

8. Bowers Jr., N., H.U. Gerber, J. Hickman, D. Jones and C. Nesbitt (1997). *Actuarial Mathematics,* 2nd Edition, Society of Actuaries, Schaumburg, IL.

9. Boyle, P. and M. Hardy (2003). Guaranteed annuity options, *Astin Bulletin,* 33, 125-152.

10. Cairns, A.J.G. (2004) *Interest Rate Models: An Introduction,* Princeton University Press, Princeton, NJ.

217

11. Cox, J., S.A. Ross and M. Rubinstein (1979). Option pricing: a simplified approach. *Journal of Financial Economics*, 7, 229-263.

12. Duffie, D. and R. Kan (1996). A yield-factor model of interest rates. *Mathematical Finance*, 6, 379-406.

13. Durrett, R. (1996). *Stochastic Calculus: A Practical Introduction*, CRC Press, New York.

14. Fabozzi, F.J.(Editor) (1997). *Handbook of Fixed Income Securities*, 5th Edition, Irwin Professional Publishing, Chicago.

15. Gelbaum, B.R. and J.M.H. Olmstead, (1964). *Counterexamples in Analysis*, Holden-Day, San Francisco.

16. Gerber, H.U. and E.S.W. Shiu (1996). Actuarial bridge to dynamic hedging and option pricing, *Insurance: Mathematics and Economics*, 18, 183-218.

17. Hardy, M. (2003). *Investment Guarantees: Modeling and Risk Management for Equity-Linked Life Insurance*, John Wiley & Sons, Toronto.

18. Harrison J.M. and S. Pliska (1981). Martingales and Stochastic integrals in the theory of continuous trading, *Stochastic Processes and Their Applications*, 11, 215-260.

19. Hassett, M. and D. Stewart (1999). *Probability for Risk Management*, ACTEX Publications, Winsted, CT.

20. Heath, D., R. Jarrow and A. Morton (1992). Bond pricing and the term structure of interest rates: a new methodology for contingent claim valuation, *Econometrica*, 60: 77-105.

21. Heath, D., R. Jarrow and A. Morton (1990). Bond pricing and the term structure of interest rates: A discrete time approximation, *Journal of Financial and Quantitative Analysis*, 25, 419-440.

22. Ho, T.S.Y. and S.B. Lee (1986). Term structure movements and pricing interest rate contingent claims, *Journal of Finance*, 41, 1011-1029.

23. Hogg, R.V. and A.T. Craig (1978). *Introduction to Mathematical Statistics*, 4th Edition, Macmillan, New York.

24. Hull, J. (1993). *Options, Futures, and Other Derivative Securities*, 2nd Edition, Prentice-Hall, Englewood Cliffs, NJ.

25. Hull, J. and A. White (1990). Pricing interest-rate derivative securities, *Review of Financial Studies*, 3, 573-592.

26. Jamshidian, F. (1991). Bond and option valuation in the Gaussian interest rate model, *Research in Finance*, 9:131-170.

27. Jamshidian, F. (1989). An exact bond option formula, *Journal of Finance*, 44, 205-209.

28. Karatzas, I. and S. Shreve (1991). *Brownian Motion and Stochastic Calculus*, 2nd Edition, Springer-Verlag, Berlin.

29. Lin, X.S. (1998). Double barrier hitting time distributions with applications to exotic options, *Insurance: Mathematics and Economics*, 23, 45-58.

30. Lin, X.S. and K.S. Tan (2003). Valuation of equity-indexed annuities under stochastic interest rates, *North American Actuarial Journal*, 7, 72-91.

31. Manistre, J. (2001). The financial economics of Universal Life: An actuarial application of stochastic calculus, preprint.

32. National Association for Variable Annuites (2004). *2004 Annuity Fact Book,* Reston, VA.

33. Øksendal, B. (1998). *Stochastic Differential Equations: An Introduction with Applications,* 5th Edition, Springer-Verlag, Berlin.

34. Pelsser, A. (2003). Pricing and hedging guaranteed annuity options via static option replication, *Insurance, Mathematics and Economics,* 33, 283-296.

35. Panjer, H.H. (Editor) (1998). *Financial Economics: with Applications to Investments, Insurance and Pensions,* Actuarial Foundation, Schaumburg, IL.

36. Ross, S.M. (1996). *Stochastic Processes,* 2nd Edition, Wiley & Sons, Inc., New York.

37. Ross, S.M. (1993). *Introduction to Probability Models,* 5th Edition, Academic Press, San Diego.

38. Rudin, W. (1976). *Principles of Mathematical Analysis,* 3rd Edition, McGraw-Hill, New York.

39. Williams, D. (1994). *Probability with Martingales,* Cambridge University Press, Cambridge, UK.

40. Vasicek, O. (1977). An equilibrium charaterization of the term structure." *Journal of Financial Economics,* 5, 177–188.

INDEX

WILEY SERIES IN PROBABILITY AND STATISTICS
ESTABLISHED BY WALTER A. SHEWHART AND SAMUEL S. WILKS

Editors: *David J. Balding, Noel A. C. Cressie, Nicholas I. Fisher,*
Iain M. Johnstone, J. B. Kadane, Geert Molenberghs. Louise M. Ryan,
David W. Scott, Adrian F. M. Smith, Jozef L. Teugels
Editors Emeriti: *Vic Barnett, J. Stuart Hunter, David G. Kendall*

The *Wiley Series in Probability and Statistics* is well established and authoritative. It covers many topics of current research interest in both pure and applied statistics and probability theory. Written by leading statisticians and institutions, the titles span both state-of-the-art developments in the field and classical methods.

Reflecting the wide range of current research in statistics, the series encompasses applied, methodological and theoretical statistics, ranging from applications and new techniques made possible by advances in computerized practice to rigorous treatment of theoretical approaches.

This series provides essential and invaluable reading for all statisticians, whether in academia, industry, government, or research.

† ABRAHAM and LEDOLTER · Statistical Methods for Forecasting
AGRESTI · Analysis of Ordinal Categorical Data
AGRESTI · An Introduction to Categorical Data Analysis
AGRESTI · Categorical Data Analysis, *Second Edition*
ALTMAN, GILL, and McDONALD · Numerical Issues in Statistical Computing for the
  Social Scientist
AMARATUNGA and CABRERA · Exploration and Analysis of DNA Microarray and
  Protein Array Data
ANDĚL · Mathematics of Chance
ANDERSON · An Introduction to Multivariate Statistical Analysis, *Third Edition*
* ANDERSON · The Statistical Analysis of Time Series
ANDERSON, AUQUIER, HAUCK, OAKES, VANDAELE, and WEISBERG ·
  Statistical Methods for Comparative Studies
ANDERSON and LOYNES · The Teaching of Practical Statistics
ARMITAGE and DAVID (editors) · Advances in Biometry
ARNOLD, BALAKRISHNAN, and NAGARAJA · Records
* ARTHANARI and DODGE · Mathematical Programming in Statistics
* BAILEY · The Elements of Stochastic Processes with Applications to the Natural
  Sciences
BALAKRISHNAN and KOUTRAS · Runs and Scans with Applications
BARNETT · Comparative Statistical Inference, *Third Edition*
BARNETT and LEWIS · Outliers in Statistical Data, *Third Edition*
BARTOSZYNSKI and NIEWIADOMSKA-BUGAJ · Probability and Statistical Inference
BASILEVSKY · Statistical Factor Analysis and Related Methods: Theory and
  Applications
BASU and RIGDON · Statistical Methods for the Reliability of Repairable Systems
BATES and WATTS · Nonlinear Regression Analysis and Its Applications
BECHHOFER, SANTNER, and GOLDSMAN · Design and Analysis of Experiments for
  Statistical Selection, Screening, and Multiple Comparisons
BELSLEY · Conditioning Diagnostics: Collinearity and Weak Data in Regression

---

*Now available in a lower priced paperback edition in the Wiley Classics Library.
†Now available in a lower priced paperback edition in the Wiley–Interscience Paperback Series.

† BELSLEY, KUH, and WELSCH · Regression Diagnostics: Identifying Influential
   Data and Sources of Collinearity
BENDAT and PIERSOL · Random Data: Analysis and Measurement Procedures,
   Third Edition
BERRY, CHALONER, and GEWEKE · Bayesian Analysis in Statistics and
   Econometrics: Essays in Honor of Arnold Zellner
BERNARDO and SMITH · Bayesian Theory
BHAT and MILLER · Elements of Applied Stochastic Processes, *Third Edition*
BHATTACHARYA and WAYMIRE · Stochastic Processes with Applications
† BIEMER, GROVES, LYBERG, MATHIOWETZ, and SUDMAN · Measurement Errors
   in Surveys
BILLINGSLEY · Convergence of Probability Measures, *Second Edition*
BILLINGSLEY · Probability and Measure, *Third Edition*
BIRKES and DODGE · Alternative Methods of Regression
BLISCHKE AND MURTHY (editors) · Case Studies in Reliability and Maintenance
BLISCHKE AND MURTHY · Reliability: Modeling, Prediction, and Optimization
BLOOMFIELD · Fourier Analysis of Time Series: An Introduction, *Second Edition*
BOLLEN · Structural Equations with Latent Variables
BOLLEN and CURRAN · Latent Curve Models: A Structural Equation Perspective
BOROVKOV · Ergodicity and Stability of Stochastic Processes
BOULEAU · Numerical Methods for Stochastic Processes
BOX · Bayesian Inference in Statistical Analysis
BOX · R. A. Fisher, the Life of a Scientist
BOX and DRAPER · Empirical Model-Building and Response Surfaces
* BOX and DRAPER · Evolutionary Operation: A Statistical Method for Process
   Improvement
BOX, HUNTER, and HUNTER · Statistics for Experimenters: Design, Innovation,
   and Discovery, *Second Editon*
BOX and LUCEÑO · Statistical Control by Monitoring and Feedback Adjustment
BRANDIMARTE · Numerical Methods in Finance: A MATLAB-Based Introduction
BROWN and HOLLANDER · Statistics: A Biomedical Introduction
BRUNNER, DOMHOF, and LANGER · Nonparametric Analysis of Longitudinal Data in
   Factorial Experiments
BUCKLEW · Large Deviation Techniques in Decision, Simulation, and Estimation
CAIROLI and DALANG · Sequential Stochastic Optimization
CASTILLO, HADI, BALAKRISHNAN, and SARABIA · Extreme Value and Related
   Models with Applications in Engineering and Science
CHAN · Time Series: Applications to Finance
CHARALAMBIDES · Combinatorial Methods in Discrete Distributions
CHATTERJEE and HADI · Sensitivity Analysis in Linear Regression
CHATTERJEE and PRICE · Regression Analysis by Example, *Third Edition*
CHERNICK · Bootstrap Methods: A Practitioner's Guide
CHERNICK and FRIIS · Introductory Biostatistics for the Health Sciences
CHILÈS and DELFINER · Geostatistics: Modeling Spatial Uncertainty
CHOW and LIU · Design and Analysis of Clinical Trials: Concepts and Methodologies,
   Second Edition
CLARKE and DISNEY · Probability and Random Processes: A First Course with
   Applications, *Second Edition*
* COCHRAN and COX · Experimental Designs, *Second Edition*
CONGDON · Applied Bayesian Modelling
CONGDON · Bayesian Statistical Modelling
CONOVER · Practical Nonparametric Statistics, *Third Edition*
COOK · Regression Graphics

*Now available in a lower priced paperback edition in the Wiley Classics Library.
†Now available in a lower priced paperback edition in the Wiley–Interscience Paperback Series.

COOK and WEISBERG · Applied Regression Including Computing and Graphics
COOK and WEISBERG · An Introduction to Regression Graphics
CORNELL · Experiments with Mixtures, Designs, Models, and the Analysis of Mixture Data, *Third Edition*
COVER and THOMAS · Elements of Information Theory
COX · A Handbook of Introductory Statistical Methods
* COX · Planning of Experiments
CRESSIE · Statistics for Spatial Data, *Revised Edition*
CSÖRGŐ and HORVÁTH · Limit Theorems in Change Point Analysis
DANIEL · Applications of Statistics to Industrial Experimentation
DANIEL · Biostatistics: A Foundation for Analysis in the Health Sciences, *Eighth Edition*
* DANIEL · Fitting Equations to Data: Computer Analysis of Multifactor Data, *Second Edition*
DASU and JOHNSON · Exploratory Data Mining and Data Cleaning
DAVID and NAGARAJA · Order Statistics, *Third Edition*
* DEGROOT, FIENBERG, and KADANE · Statistics and the Law
DEL CASTILLO · Statistical Process Adjustment for Quality Control
DeMARIS · Regression with Social Data: Modeling Continuous and Limited Response Variables
DEMIDENKO · Mixed Models: Theory and Applications
DENISON, HOLMES, MALLICK and SMITH · Bayesian Methods for Nonlinear Classification and Regression
DETTE and STUDDEN · The Theory of Canonical Moments with Applications in Statistics, Probability, and Analysis
DEY and MUKERJEE · Fractional Factorial Plans
DILLON and GOLDSTEIN · Multivariate Analysis: Methods and Applications
DODGE · Alternative Methods of Regression
* DODGE and ROMIG · Sampling Inspection Tables, *Second Edition*
* DOOB · Stochastic Processes
DOWDY, WEARDEN, and CHILKO · Statistics for Research, *Third Edition*
DRAPER and SMITH · Applied Regression Analysis, *Third Edition*
DRYDEN and MARDIA · Statistical Shape Analysis
DUDEWICZ and MISHRA · Modern Mathematical Statistics
DUNN and CLARK · Basic Statistics: A Primer for the Biomedical Sciences, *Third Edition*
DUPUIS and ELLIS · A Weak Convergence Approach to the Theory of Large Deviations
* ELANDT-JOHNSON and JOHNSON · Survival Models and Data Analysis
† ENDERS · Applied Econometric Time Series
† ETHIER and KURTZ · Markov Processes: Characterization and Convergence
EVANS, HASTINGS, and PEACOCK · Statistical Distributions, *Third Edition*
FELLER · An Introduction to Probability Theory and Its Applications, Volume I, *Third Edition,* Revised; Volume II, *Second Edition*
FISHER and VAN BELLE · Biostatistics: A Methodology for the Health Sciences
FITZMAURICE, LAIRD, and WARE · Applied Longitudinal Analysis
* FLEISS · The Design and Analysis of Clinical Experiments
FLEISS · Statistical Methods for Rates and Proportions, *Third Edition*
† FLEMING and HARRINGTON · Counting Processes and Survival Analysis
FULLER · Introduction to Statistical Time Series, *Second Edition*
FULLER · Measurement Error Models
GALLANT · Nonlinear Statistical Models
GEISSER · Modes of Parametric Statistical Inference
GEWEKE · Contemporary Bayesian Econometrics and Statistics

---

*Now available in a lower priced paperback edition in the Wiley Classics Library.
†Now available in a lower priced paperback edition in the Wiley–Interscience Paperback Series.

GHOSH, MUKHOPADHYAY, and SEN · Sequential Estimation
GIESBRECHT and GUMPERTZ · Planning, Construction, and Statistical Analysis of
   Comparative Experiments
GIFI · Nonlinear Multivariate Analysis
GIVENS and HOETING · Computational Statistics
GLASSERMAN and YAO · Monotone Structure in Discrete-Event Systems
GNANADESIKAN · Methods for Statistical Data Analysis of Multivariate Observations,
   Second Edition
GOLDSTEIN and LEWIS · Assessment: Problems, Development, and Statistical Issues
GREENWOOD and NIKULIN · A Guide to Chi-Squared Testing
GROSS and HARRIS · Fundamentals of Queueing Theory, *Third Edition*
† GROVES · Survey Errors and Survey Costs
* HAHN and SHAPIRO · Statistical Models in Engineering
HAHN and MEEKER · Statistical Intervals: A Guide for Practitioners
HALD · A History of Probability and Statistics and their Applications Before 1750
HALD · A History of Mathematical Statistics from 1750 to 1930
† HAMPEL · Robust Statistics: The Approach Based on Influence Functions
HANNAN and DEISTLER · The Statistical Theory of Linear Systems
HEIBERGER · Computation for the Analysis of Designed Experiments
HEDAYAT and SINHA · Design and Inference in Finite Population Sampling
HELLER · MACSYMA for Statisticians
HINKELMANN and KEMPTHORNE · Design and Analysis of Experiments, Volume 1:
   Introduction to Experimental Design
HINKELMANN and KEMPTHORNE · Design and Analysis of Experiments, Volume 2:
   Advanced Experimental Design
HOAGLIN, MOSTELLER, and TUKEY · Exploratory Approach to Analysis
   of Variance
HOAGLIN, MOSTELLER, and TUKEY · Exploring Data Tables, Trends and Shapes
* HOAGLIN, MOSTELLER, and TUKEY · Understanding Robust and Exploratory
   Data Analysis
HOCHBERG and TAMHANE · Multiple Comparison Procedures
HOCKING · Methods and Applications of Linear Models: Regression and the Analysis
   of Variance, *Second Edition*
HOEL · Introduction to Mathematical Statistics, *Fifth Edition*
HOGG and KLUGMAN · Loss Distributions
HOLLANDER and WOLFE · Nonparametric Statistical Methods, *Second Edition*
HOSMER and LEMESHOW · Applied Logistic Regression, *Second Edition*
HOSMER and LEMESHOW · Applied Survival Analysis: Regression Modeling of
   Time to Event Data
† HUBER · Robust Statistics
HUBERTY · Applied Discriminant Analysis
HUNT and KENNEDY · Financial Derivatives in Theory and Practice
HUSKOVA, BERAN, and DUPAC · Collected Works of Jaroslav Hajek—
   with Commentary
HUZURBAZAR · Flowgraph Models for Multistate Time-to-Event Data
IMAN and CONOVER · A Modern Approach to Statistics
† JACKSON · A User's Guide to Principle Components
JOHN · Statistical Methods in Engineering and Quality Assurance
JOHNSON · Multivariate Statistical Simulation
JOHNSON and BALAKRISHNAN · Advances in the Theory and Practice of Statistics: A
   Volume in Honor of Samuel Kotz
JOHNSON and BHATTACHARYYA · Statistics: Principles and Methods, *Fifth Edition*

*Now available in a lower priced paperback edition in the Wiley Classics Library.
†Now available in a lower priced paperback edition in the Wiley–Interscience Paperback Series.

JOHNSON and KOTZ · Distributions in Statistics

JOHNSON and KOTZ (editors) · Leading Personalities in Statistical Sciences: From the Seventeenth Century to the Present

JOHNSON, KOTZ, and BALAKRISHNAN · Continuous Univariate Distributions, Volume 1, *Second Edition*

JOHNSON, KOTZ, and BALAKRISHNAN · Continuous Univariate Distributions, Volume 2, *Second Edition*

JOHNSON, KOTZ, and BALAKRISHNAN · Discrete Multivariate Distributions

JOHNSON, KEMP, and KOTZ · Univariate Discrete Distributions, *Third Edition*

JUDGE, GRIFFITHS, HILL, LÜTKEPOHL, and LEE · The Theory and Practice of Econometrics, *Second Edition*

JUREČKOVÁ and SEN · Robust Statistical Procedures: Aymptotics and Interrelations

JUREK and MASON · Operator-Limit Distributions in Probability Theory

KADANE · Bayesian Methods and Ethics in a Clinical Trial Design

KADANE AND SCHUM · A Probabilistic Analysis of the Sacco and Vanzetti Evidence

KALBFLEISCH and PRENTICE · The Statistical Analysis of Failure Time Data, *Second Edition*

KASS and VOS · Geometrical Foundations of Asymptotic Inference

† KAUFMAN and ROUSSEEUW · Finding Groups in Data: An Introduction to Cluster Analysis

KEDEM and FOKIANOS · Regression Models for Time Series Analysis

KENDALL, BARDEN, CARNE, and LE · Shape and Shape Theory

KHURI · Advanced Calculus with Applications in Statistics, *Second Edition*

KHURI, MATHEW, and SINHA · Statistical Tests for Mixed Linear Models

* KISH · Statistical Design for Research

KLEIBER and KOTZ · Statistical Size Distributions in Economics and Actuarial Sciences

KLUGMAN, PANJER, and WILLMOT · Loss Models: From Data to Decisions, *Second Edition*

KLUGMAN, PANJER, and WILLMOT · Solutions Manual to Accompany Loss Models: From Data to Decisions, *Second Edition*

KOTZ, BALAKRISHNAN, and JOHNSON · Continuous Multivariate Distributions, Volume 1, *Second Edition*

KOTZ and JOHNSON (editors) · Encyclopedia of Statistical Sciences: Volumes 1 to 9 with Index

KOTZ and JOHNSON (editors) · Encyclopedia of Statistical Sciences: Supplement Volume

KOTZ, READ, and BANKS (editors) · Encyclopedia of Statistical Sciences: Update Volume 1

KOTZ, READ, and BANKS (editors) · Encyclopedia of Statistical Sciences: Update Volume 2

KOVALENKO, KUZNETZOV, and PEGG · Mathematical Theory of Reliability of Time-Dependent Systems with Practical Applications

LACHIN · Biostatistical Methods: The Assessment of Relative Risks

LAD · Operational Subjective Statistical Methods: A Mathematical, Philosophical, and Historical Introduction

LAMPERTI · Probability: A Survey of the Mathematical Theory, *Second Edition*

LANGE, RYAN, BILLARD, BRILLINGER, CONQUEST, and GREENHOUSE · Case Studies in Biometry

LARSON · Introduction to Probability Theory and Statistical Inference, *Third Edition*

LAWLESS · Statistical Models and Methods for Lifetime Data, *Second Edition*

LAWSON · Statistical Methods in Spatial Epidemiology

LE · Applied Categorical Data Analysis

LE · Applied Survival Analysis

LEE and WANG · Statistical Methods for Survival Data Analysis, *Third Edition*

---

*Now available in a lower priced paperback edition in the Wiley Classics Library.

†Now available in a lower priced paperback edition in the Wiley–Interscience Paperback Series.

LePAGE and BILLARD · Exploring the Limits of Bootstrap
LEYLAND and GOLDSTEIN (editors) · Multilevel Modelling of Health Statistics
LIAO · Statistical Group Comparison
LINDVALL · Lectures on the Coupling Method
LIN · Introductory Stochastic Analysis for Finance and Insurance
LINHART and ZUCCHINI · Model Selection
LITTLE and RUBIN · Statistical Analysis with Missing Data, *Second Edition*
LLOYD · The Statistical Analysis of Categorical Data
LOWEN and TEICH · Fractal-Based Point Processes
MAGNUS and NEUDECKER · Matrix Differential Calculus with Applications in
    Statistics and Econometrics, *Revised Edition*
MALLER and ZHOU · Survival Analysis with Long Term Survivors
MALLOWS · Design, Data, and Analysis by Some Friends of Cuthbert Daniel
MANN, SCHAFER, and SINGPURWALLA · Methods for Statistical Analysis of
    Reliability and Life Data
MANTON, WOODBURY, and TOLLEY · Statistical Applications Using Fuzzy Sets
MARCHETTE · Random Graphs for Statistical Pattern Recognition
MARDIA and JUPP · Directional Statistics
MASON, GUNST, and HESS · Statistical Design and Analysis of Experiments with
    Applications to Engineering and Science, *Second Edition*
McCULLOCH and SEARLE · Generalized, Linear, and Mixed Models
McFADDEN · Management of Data in Clinical Trials
* McLACHLAN · Discriminant Analysis and Statistical Pattern Recognition
McLACHLAN, DO, and AMBROISE · Analyzing Microarray Gene Expression Data
McLACHLAN and KRISHNAN · The EM Algorithm and Extensions
McLACHLAN and PEEL · Finite Mixture Models
McNEIL · Epidemiological Research Methods
MEEKER and ESCOBAR · Statistical Methods for Reliability Data
MEERSCHAERT and SCHEFFLER · Limit Distributions for Sums of Independent
    Random Vectors: Heavy Tails in Theory and Practice
MICKEY, DUNN, and CLARK · Applied Statistics: Analysis of Variance and
    Regression, *Third Edition*
* MILLER · Survival Analysis, *Second Edition*
MONTGOMERY, PECK, and VINING · Introduction to Linear Regression Analysis,
    Third Edition
MORGENTHALER and TUKEY · Configural Polysampling: A Route to Practical
    Robustness
MUIRHEAD · Aspects of Multivariate Statistical Theory
MULLER and STOYAN · Comparison Methods for Stochastic Models and Risks
MURRAY · X-STAT 2.0 Statistical Experimentation, Design Data Analysis, and
    Nonlinear Optimization
MURTHY, XIE, and JIANG · Weibull Models
MYERS and MONTGOMERY · Response Surface Methodology: Process and Product
    Optimization Using Designed Experiments, *Second Edition*
MYERS, MONTGOMERY, and VINING · Generalized Linear Models. With
    Applications in Engineering and the Sciences
† NELSON · Accelerated Testing, Statistical Models, Test Plans, and Data Analyses
† NELSON · Applied Life Data Analysis
NEWMAN · Biostatistical Methods in Epidemiology
OCHI · Applied Probability and Stochastic Processes in Engineering and Physical
    Sciences
OKABE, BOOTS, SUGIHARA, and CHIU · Spatial Tesselations: Concepts and
    Applications of Voronoi Diagrams, *Second Edition*
OLIVER and SMITH · Influence Diagrams, Belief Nets and Decision Analysis

*Now available in a lower priced paperback edition in the Wiley Classics Library.
†Now available in a lower priced paperback edition in the Wiley–Interscience Paperback Series.

PALTA · Quantitative Methods in Population Health: Extensions of Ordinary Regressions
PANKRATZ · Forecasting with Dynamic Regression Models
PANKRATZ · Forecasting with Univariate Box-Jenkins Models: Concepts and Cases
* PARZEN · Modern Probability Theory and Its Applications
PEÑA, TIAO, and TSAY · A Course in Time Series Analysis
PIANTADOSI · Clinical Trials: A Methodologic Perspective
PORT · Theoretical Probability for Applications
POURAHMADI · Foundations of Time Series Analysis and Prediction Theory
PRESS · Bayesian Statistics: Principles, Models, and Applications
PRESS · Subjective and Objective Bayesian Statistics, *Second Edition*
PRESS and TANUR · The Subjectivity of Scientists and the Bayesian Approach
PUKELSHEIM · Optimal Experimental Design
PURI, VILAPLANA, and WERTZ · New Perspectives in Theoretical and Applied
  Statistics
† PUTERMAN · Markov Decision Processes: Discrete Stochastic Dynamic Programming
QIU · Image Processing and Jump Regression Analysis
* RAO · Linear Statistical Inference and Its Applications, *Second Edition*
RAUSAND and HØYLAND · System Reliability Theory: Models, Statistical Methods,
  and Applications, *Second Edition*
RENCHER · Linear Models in Statistics
RENCHER · Methods of Multivariate Analysis, *Second Edition*
RENCHER · Multivariate Statistical Inference with Applications
* RIPLEY · Spatial Statistics
RIPLEY · Stochastic Simulation
ROBINSON · Practical Strategies for Experimenting
ROHATGI and SALEH · An Introduction to Probability and Statistics, *Second Edition*
ROLSKI, SCHMIDLI, SCHMIDT, and TEUGELS · Stochastic Processes for Insurance
  and Finance
ROSENBERGER and LACHIN · Randomization in Clinical Trials: Theory and Practice
ROSS · Introduction to Probability and Statistics for Engineers and Scientists
† ROUSSEEUW and LEROY · Robust Regression and Outlier Detection
* RUBIN · Multiple Imputation for Nonresponse in Surveys
RUBINSTEIN · Simulation and the Monte Carlo Method
RUBINSTEIN and MELAMED · Modern Simulation and Modeling
RYAN · Modern Regression Methods
RYAN · Statistical Methods for Quality Improvement, *Second Edition*
SALEH · Theory of Preliminary Test and Stein-Type Estimation with Applications
* SCHEFFE · The Analysis of Variance
SCHIMEK · Smoothing and Regression: Approaches, Computation, and Application
SCHOTT · Matrix Analysis for Statistics, *Second Edition*
SCHOUTENS · Levy Processes in Finance: Pricing Financial Derivatives
SCHUSS · Theory and Applications of Stochastic Differential Equations
SCOTT · Multivariate Density Estimation: Theory, Practice, and Visualization
SEARLE · Linear Models for Unbalanced Data
SEARLE · Matrix Algebra Useful for Statistics
SEARLE, CASELLA, and McCULLOCH · Variance Components
SEARLE and WILLETT · Matrix Algebra for Applied Economics
SEBER and LEE · Linear Regression Analysis, *Second Edition*
† SEBER · Multivariate Observations
† SEBER and WILD · Nonlinear Regression
SENNOTT · Stochastic Dynamic Programming and the Control of Queueing Systems
* SERFLING · Approximation Theorems of Mathematical Statistics
SHAFER and VOVK · Probability and Finance: It's Only a Game!

---

*Now available in a lower priced paperback edition in the Wiley Classics Library.
†Now available in a lower priced paperback edition in the Wiley–Interscience Paperback Series.

SILVAPULLE and SEN · Constrained Statistical Inference: Inequality, Order, and Shape Restrictions

SMALL and McLEISH · Hilbert Space Methods in Probability and Statistical Inference

SRIVASTAVA · Methods of Multivariate Statistics

STAPLETON · Linear Statistical Models

STAUDTE and SHEATHER · Robust Estimation and Testing

STOYAN, KENDALL, and MECKE · Stochastic Geometry and Its Applications, *Second Edition*

STOYAN and STOYAN · Fractals, Random Shapes and Point Fields: Methods of Geometrical Statistics

STYAN · The Collected Papers of T. W. Anderson: 1943–1985

SUTTON, ABRAMS, JONES, SHELDON, and SONG · Methods for Meta-Analysis in Medical Research

TAKEZAWA · Introduction to Nonparametric Regression

TANAKA · Time Series Analysis: Nonstationary and Noninvertible Distribution Theory

THOMPSON · Empirical Model Building

THOMPSON · Sampling, *Second Edition*

THOMPSON · Simulation: A Modeler's Approach

THOMPSON and SEBER · Adaptive Sampling

THOMPSON, WILLIAMS, and FINDLAY · Models for Investors in Real World Markets

TIAO, BISGAARD, HILL, PEÑA, and STIGLER (editors) · Box on Quality and Discovery: with Design, Control, and Robustness

TIERNEY · LISP-STAT: An Object-Oriented Environment for Statistical Computing and Dynamic Graphics

TSAY · Analysis of Financial Time Series, *Second Edition*

UPTON and FINGLETON · Spatial Data Analysis by Example, Volume II: Categorical and Directional Data

VAN BELLE · Statistical Rules of Thumb

VAN BELLE, FISHER, HEAGERTY, and LUMLEY · Biostatistics: A Methodology for the Health Sciences, *Second Edition*

VESTRUP · The Theory of Measures and Integration

VIDAKOVIC · Statistical Modeling by Wavelets

VINOD and REAGLE · Preparing for the Worst: Incorporating Downside Risk in Stock Market Investments

WALLER and GOTWAY · Applied Spatial Statistics for Public Health Data

WEERAHANDI · Generalized Inference in Repeated Measures: Exact Methods in MANOVA and Mixed Models

WEISBERG · Applied Linear Regression, *Third Edition*

WELSH · Aspects of Statistical Inference

WESTFALL and YOUNG · Resampling-Based Multiple Testing: Examples and Methods for *p*-Value Adjustment

WHITTAKER · Graphical Models in Applied Multivariate Statistics

WINKER · Optimization Heuristics in Economics: Applications of Threshold Accepting

WONNACOTT and WONNACOTT · Econometrics, *Second Edition*

WOODING · Planning Pharmaceutical Clinical Trials: Basic Statistical Principles

WOODWORTH · Biostatistics: A Bayesian Introduction

WOOLSON and CLARKE · Statistical Methods for the Analysis of Biomedical Data, *Second Edition*

WU and HAMADA · Experiments: Planning, Analysis, and Parameter Design Optimization

YANG · The Construction Theory of Denumerable Markov Processes

* ZELLNER · An Introduction to Bayesian Inference in Econometrics

ZHOU, OBUCHOWSKI, and McCLISH · Statistical Methods in Diagnostic Medicine

*Now available in a lower priced paperback edition in the Wiley Classics Library.

†Now available in a lower priced paperback edition in the Wiley–Interscience Paperback Series.